# Design for Assembly

A Monograph covering:
- Process Definition & Part Sequencing
- Product Development Guidelines
- Part Design for Feeding, Insertion and Fastening
- Product Assembly Improvement Process
- Quantifying Ease of Assembly

Carl F. Zorowski

Design for Assembly

Copyright 2016
All Rights Reserved

Design for Assembly

"Scientists discover the world that exists; engineers create the world that never was."

- Theodore von Karman
  mechanical engineer

# Design for Assembly

Design for Assembly

# Table of Contents

## Chapter 1 – Process Definition and Part Sequencing
- Potential DFA Impact — 1
- Reasons for Neglect — 2
- Assembly Process Definition — 2
- Assembly Functions — 3
- Handling — 4
- Composition — 5
- Composition Examples — 6
- Checking — 7
- Assembly Process Tree — 7
- Composition Demonstration — 8
- Assembly Objects — 11
- Reasons for Assembly — 12
- Stopper Valve Example — 13
- Part Sequence Diagram — 17
- Toggle Switch Example — 18

## Chapter 2 – Product Development Guidelines
- Principles of Product Design — 27
- Importance of Base — 28
- Layered Assembly — 30
- Modular Components — 32
- Accommodation for Errors — 33
- Fastener Reduction — 35
- Minimizing Parts — 39
- Flexible Items — 43
- Part Redundancy — 44

## Chapter 3 – Part Design for Feeding and Insertion

- Four Part Spindle Subassembly 47
- Applying the Loctite 47
- Gasket Addition 48
- Screw Insertion and Adjustment 49
- Assembly Sequence Diagram 49
- Part Design Guidelines 50
- Tangling, Overlapping and Nesting 51
- Protrusion Examples 52
- Tangling Examples 54
- Overlapping Example 55
- Gravity Examples 56
- Overlooked Examples 57
- Nesting and Clinging Example 58
- Symmetry 59
- Symmetry Examples 60
- Asymmetry 62
- Asymmetry Examples 63

## Chapter 4 – Product Assembly Improvement Process

- Product Redesign Process 65
- Damper Valve 66
  - Part Sequence Diagram 67
  - DFA Systems Analysis 67
  - Design Change Candidates 68
- Redesigned Damper Valve 69
- Pneumatic Pressure Sensor 70

Design for Assembly

|  |  |
|---|---|
| Part Sequence Diagram | 71 |
| DFA Systems Analysis | 72 |
| Design Change Candidates | 73 |
| • Redesigned Pressure Sensor | 73 |
| • Light Switch Redesign | 74 |
| • Plastic Pipe Cutter Redesign | 76 |
| • Copier Latch Subassembly | 78 |

## Chapter 5 – Quantify Ease of Assembly

- Basis of Quantifying Tool — 81
- Part Merit Rating — 82
- Geometric Interpretation — 82
- Part Merit Calculation — 83
- Assembly Merit — 83
- Feeding Ratings — 84
- Insertion Ratings — 85
- Fastener Time Comparisons — 86
- Pneumatic Pressure Sensor — 87
- Original Pneumatic Sensor Rating — 87
- Redesigned Pneumatic Sensor Rating — 92
- Pipe Cutter Redesign — 95
- Pipe Cutter CAM and PAM — 95
- Comparison of results — 96

# Design for Assembly

# Preface

New product development always involves a large number of competing requirement specifications. There can be hundreds of factors dealing with functionality, manufacturing, safety, marketing, maintenance and product retirement, etc. that require attention in the product development process. With an understandable high early priority on functional design and economic manufacturability a consideration often neglected or overlooked is the impact that Design for Assembly, DFA, can have on cost reduction and performance improvement of the final product.

This monograph is a collection of suggestions, guidelines and examples that highlight how DFA can benefit designers in developing a product that more nearly achieves desired "ideal" outcomes. Although the application of DFA is most effective in the conceptual phase of a new product design it is best to learn about its principles and use by examining the design of existing products and their potential improvement. This is the method used in this monograph. Following is a brief discussion of the material covered in the five chapters of this book.

Chapter-1 deals with the importance of DFA, an operational definition of assembly, part characterization, reasons for separate parts and creating a apart sequence diagram.

# Design for Assembly

Chapter-2 presents and discusses with examples generally accepted DFA product development guidelines. These include: providing a base for assembly, layering and stacking components, using multi-functional parts, accommodating for errors, reducing fasteners, limiting flexible items and minimizing part count.

Chapter-3 discusses the design of parts to facilitate their feeding and insertion. Topics include the avoidance of tangling, overlapping and nesting, the role of gravity, the value of geometric symmetry and the use of asymmetry.

Chapter-4 presents a four-step process for applying DFA principles to the improvement of existing products. A number of examples of the process application are included and discussed in detail.

Chapter-5 introduces a subjective technique that numerically quantifies the ease of assembly based on the feeding, insertion and fastening of parts. It is used to quantitatively compare the degree of improvement that DFA makes in a number of product examples.

The underlying philosophy of DFA is the pursuit of simplicity and elegance in the assembly of the physical design while insuring that the functionality and manufacturability goals are met. This requires the designer to employ skills of creativity and inventiveness in contrast to the analytic

Design for Assembly

and synthesis skills development that is emphasized in the formal education of engineers.

The material contained in this monograph are from notes in a Mechanical Design Engineering course taught at North Carolina State University by the author. An audio supplemented version of the subject is available at
 *http://www.designengineeringreview.com.*
A full course supplement is also available at
 *http://www.designforassembly.com.*

Carl F. Zorowski
Cary, NC
October 2016

# Design for Assembly

Design for Assembly

# Chapter 1 – Process Definition and Part Sequencing

This chapter deals with an introduction to DFA as an important part of the total product design process. It includes its potential impact on product cost, why it is sometimes neglected, an operational definition of the assembly process, reasons for assembly and the value of part sequencing in understanding assembly.

## Potential DFA Impact

Applying the principles of product and part design can significantly reduce the total cost of production of a device. This is accomplished by a reduction in assembly time, the need for fewer subassemblies, smaller parts inventories, simpler assembly systems and reduced manufacturing costs. All of these contribute to a higher quality final product.

- reduction in assembly time
- fewer subassemblies
- smaller part inventories
- reduced manufacturing costs
- simpler assembly systems
- higher quality final products

*Figure 1-1: Impact of DFA Application*

Unfortunately DFA is an area of product development and design that often receives little attention or may even be neglected.

## Reasons for Neglect

Design for Assembly can often be short-changed as a consequence of existing design practices and assembly and manufacturing cost considerations. In many instances product function dominates early design decisions. Established traditional design practices can be difficult to overcome. Cost considerations dealing with part manufacturing or the assembly system required by the product are also high priority early design issues. Inclusion of Design for Assembly is often relegated to a minor role ad doesn't receive the priority it deserves.

- product function dominates early design
- traditional practices difficult to overcome
- part manufacturing dominate cost concerns
- product design dictates assembly system
- ➡ assembly process not well understood
- ➡ product and part DFA principles unknown
- ➡ measuring DFA impact is at best subjective

*Figure 1-2: Why DFA is Neglected*

DFA may be neglected because the assembly process as a physical operation is not well understood. DFA principles of part and product design may be unknown and measuring the impact of DFA in the initial design development process is difficult to measure objectively. These are the issues addressed by this monograph.

## Assembly Process Definition

The definition of assembly is best understood from an operational perspective. It is the act of bringing together parts, components, materials or subassemblies in alignment or spatial relationship to one another and providing them with a mechanism to maintain that relationship by some method of joining, fastening or containing against external effects to produce some functional complex product.

**The main purpose of "assembly" is to**
bring together
**components, materials and subassemblies**
**to be** properly aligned
**and**
joined or fastened
**into functional complex products**

*Figure 1-3: Purpose of Assembly*

## Assembly Functions

This operational definition leads to three functions that define the process. These are handling, composing, and checking. The handling function deals with the activity required to bring two parts together into a specific spatial relationship or alignment. Composing includes those activities that ensure the prescribed alignment of these parts or components can be maintained against outside effects that attempt to disrupt that relationship. Finally, checking includes those activities that determine that the first two functions have been properly performed. At this point assembly has been accomplished.

# Design for Assembly

### Handling
- placing two or more objects into a particular position

### Composing (Joining or Fastening)
- ensuring mutual relationship against outside effects

### Checking
- determining that the first two functions have been carried out properly

*Figure 1-4: Assembly Functions*

## Handling

Handling is further subdivided into three sub functions or activities. These are storing, transporting and positioning. Before parts can be assembled they must be stored in some fashion. They might be simply

### Storing
- boxes, hoppers, magazines, racks, pallets, bins

### Transporting
- moving, separating, merging, orientation turning, rotating, allocating

### Positioning
- locating, aligning, inserting

*Figure 1-5: Handling Function*

put into boxes or placed on pallets or more carefully arranged in hoppers, magazines or on racks. These stored parts must then be transported to where the assembly will take place. In addition to their being physically moved they may also need to be separated, oriented and allocated for the next step. Positioning is where parts are brought into proper physical alignment with respect to each other in preparation for the next major function, composing.

## Composition

Composing, often thought of as the act of joining and fastening, is the assembly function that provides the mechanism by which parts maintain their physical spatial relationship with one another against outside effects that attempt to disrupt that relationship. It is best understood in terms of the means and mechanisms used to achieve the composition. The means in a welded joint

**Means**
  - force, form, material

**Mechanisms**
  - joining, filling, interference, phase change, change of form

*Figure 1-6: Means & Mechanisms*

can be thought of as the heating of the welding rod with the mechanism being a change in phase from solid to molten metal both in the rod and adjoining parts fusing all three together into a solid continuous joint when solidification take place.

## Composition Examples

Figure 1-7 includes a number of examples of composition illustrating that parts can be held together by means other than nuts and bolts. The first example of welding has already been discussed. In the second example composition takes place when a gas is placed in a tank under pressure. The means is the shape and containment of the tank that restricts the diffusion of the gas. The mechanism is the use of a pressure differential to fill the tank with gas. The third example is a shrink fit were part interference is used to create a friction force that restricts relative motion between the parts. When two objects are glued together the means becomes the adhesive force that holds the part together generated

| **Composing** | | |
|---|---|---|
| Example | Means (force, form, material) | Mechanisms (how means are achieved) |
| welding of plates | welding rod | change of phase |
| placing gas in tank | shape of tank | filling |
| shrink fit | friction force | part interference |
| glueing | adhesive force | chemical change |
| rivet plates | clamping force | deform metal |
| geo mag toy | magnetic force | magnetized metal |

*Figure 1-7: Examples of Composition*

when the glue undergoes chemical change. In riveting two plates together the means is a clamping force created when the metal of the rivet is deformed. Magnetic forces used to hold parts together by bringing properly aligned magnetized materials together is another example of composition. As stated previously, composing provides the means and mechanism by which parts maintain their physical spatial relationship with one another against outside effects that attempt to disrupt that relationship.

## Checking

Checking in the assembly process deals with determining if the first two functions of handling and composing were completed correctly. It consists of three sub-functions: presence, position and quality of composition. Presence asks the question is the part located where it should be. Position asks the question is the alignment of the part correct for its final function. Finally, quality of composition asks the question can the part properly withstand the external effects that could

**Presence**
- is part in place?

**Position**
- is alignment correct?

**Quality of Composition**
- can external effects be withstood?

*Figure 1-8: Checking of Assembly*

disrupt it proper function in the product. If all of these questions can be answered affirmatively then successful assembly has taken place.

## Assembly Process Tree

The entire assembly process as operationally defined is depicted in the tree diagram in Figure 1-9 with each of the major assembly functions of handling, composing and checking and their sub-functions indicated.

# Design for Assembly

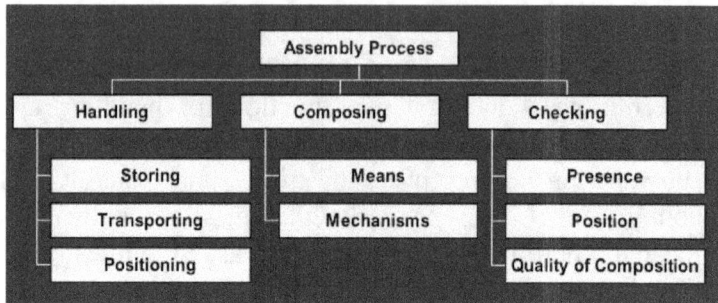

*Figure 1-9: Assembly Process*

In even the simplest of assembly operations all of these functions and sub-functions are involved even if there isn't a conscious awareness of them taking place.

## Composition Demonstration

The simple act of stapling two pages together to produce a document illustrates all steps of the assembly process.

The handling function requires obtaining the two pages from their original stored location and transporting them into proximity with each other. (Figure 1-10) They must then be positioned properly in preparation for composition.

*Figure 1-10: A lightening the Pages*

Design for Assembly

In composing the aligned sheets are first placed between the jaws of the stapler to the corner where attachment will take place. (Figure 1 – 11)

*Figure 1-11: Preparing for Composition*

Force is then applied to push the metal staple through the sheets of paper and against the anvil that bends and closes the staple (Figure 1-12). This completes the composition.

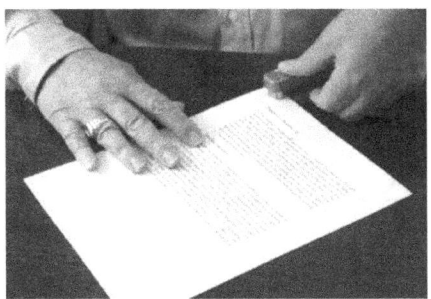

*Figure 1-12: Act of Stapling*

In the checking phase the first step is to determine if a staple has been inserted. The second step is to check if it is in the proper location. (Figure 1-13)

*Figure 1-13: Checking Location*

*Figure 1-14: Quality of Staple*

Finally is the quality of the clinch sufficient to keep the two papers together as a functional document?

## Assembly Objects

Individual objects or collections of parts that undergo assembly are classified and described primarily by their function. An individual part composed of a single material is designated a *machine part*. Assembled collections of parts that are some portion of a more complex assembly are referred to as *subassemblies*. A *building block* is a subassembly that meets specific

# Design for Assembly

assembly or functional requirements. (An alternator on an automotive engine would fit this category.)

- **Machine Part** — individual part, composed of a single material
- **Subassembly** – an assembled collection of parts
- **Building Block** – subassembly that meets specific assembly or functional requirements
- **Component** (generic term) - machine part, subassembly or building block
- **Base Component** – a (larger) component on to which other components can be placed
- **Formless Material** – glue, paint, liquids, etc.

*Figure 1-15: Assembly Objects*

A *component* is a generic term that can be applied to any of the assembly objects. A *base component* is usually a larger component on to which other components are placed and composed. Materials like glues, paints, liquids, lubricants, etc. are referred to as *formless material*.

The universal rod end illustrated in Figure 1-16 is composed of a number of these defined assembly objects. The rod end body represents a base component into which other components will be inserted. The washers and locking rings together with the ball end represents components but can also be classified as machine parts. When composition of all the parts is completed the final product can be classified as a building block.

Design for Assembly

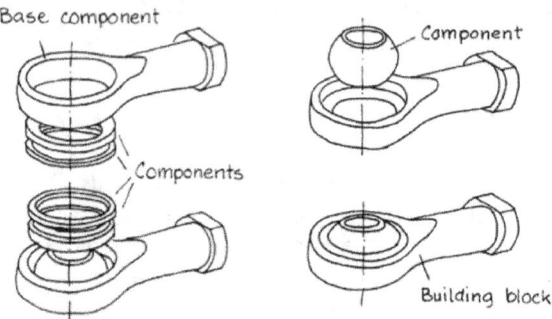

*Figure 1-16: Assembly of a Universal Rod*

Reasons For Assembly

This section deals with why separate parts are needed. Although not totally comprehensive this list covers the most important reasons. If two parts are required to move relative to one another in general they will have to be separate parts. Material differentiation is another reason. Individual conductive parts that may need to be separated by an insulating component illustrate this. Sometimes it is less costly to manufacture two parts and fasten them together rather than manufacturing a single part that is more costly or

- Movement
- Material Differentiation
- Production Considerations
- Replacability
- Functional Requirement
- Aesthetic Considerations

*Figure 1-17: Need for Separate Parts*

wasteful of material. This reason is classified as a production consideration. If parts wear out through use they need to be replaced, hence a separate part is required. Gaskets and seals are examples. A special functional requirement within a design may only be satisfied by the creation of a separate part. Finally, aesthetic considerations may dictate the need for separate parts that don't really contribute to the functionality. Trim elements or logo decals are examples.

## Stopper Valve Example

A simple stopper valve (Figure 1-18) will be used to illustrate the need and reasons for separate parts. The valve consists of a vertical tubular stem that holds an annular semi spherical rubber stopper supported by a circular ring and drift pin against a circular plate. The stem is moved vertically by a clevis mechanism attached at its top. The valve rests when closed against a beveled circular hole in a plate welded to a short hollow tube that protects the bottom of the valve stem. The entire device is fastened to a hole in a tank permitting easy removal for maintenance.

A number of parts (1 through 5) are identified on the device. Consider which of the six reasons for separate parts listed in Figure 1-17 apply to these five parts.

Design for Assembly

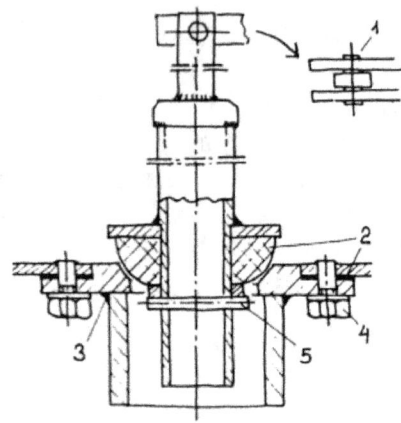

*Figure 1-18: Stopper Valve*

Movement is reason for the parts in area 1 (Figure 1-19). The parts in the clevis need to move relative to one another for it to operate properly as the stem is moved up and down.

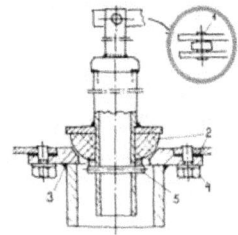

**1. Movement**

*Figure 1-19: Movement*

The rubber topper and the gasket in area 2, (Figure 1- 20), must both be made of flexible materials to function properly. This is a good example of material differentiation as a reason for separate parts.

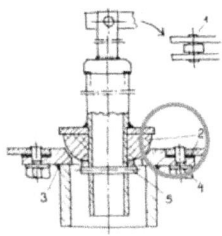

1. Movement
2. Material Differentiation

*Figure 1-20: Material Differentiation*

Production considerations are the reason for the mounting plate and stem protector to be separate in area 3, (Figure 1 – 21). It would be too costly to machine a single part as a replacement.

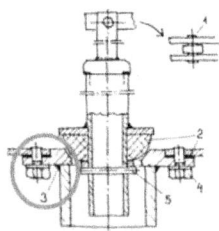

1. Movement
2. Material Differentiation
3. Production Considerations

*Figure 1-21: Production Considerations*

Separate fasteners in area 4 are required to attach the valve to the tank in order to permit its removal for replacement of the rubber stopper when it becomes worn.

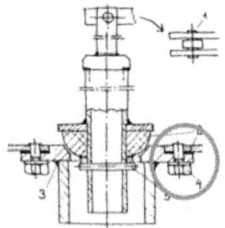

1. Movement
2. Material Differentiation
3. Production Considerations
4. Replaceable

*Figure 1-22: Replacement*

The steel ring and drift pin in area 5, (figure 1 – 23) are required to be separate to satisfy the functional requirement of holding the rubber topper in place properly and with sufficient force in this composition.

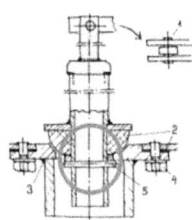

1. Movement
2. Material Differentiation
3. Production Considerations
4. Replaceable
5. Functional Requirement

*Figure 23: Functional Requirement*

Design for Assembly

## Part Sequence Diagram

Product specifications contain a bill of material listing all the parts that make up the final assembly. These are often included with the assembly drawings of the product. Although the bill of materials lists all the parts required it does not indicate the order in which these parts should be put together.

In applying the principles of design for assembly it is very helpful to prepare a part sequence diagram indicating the order in which the parts are inserted into the product as it is assembled. The creation of a part sequence diagram will be illustrated using the simple single pole toggle switch shown in Figure 1 - 24.

*Figure 1-24: Toggle Switch*

## Toggle Switch Example

In this example, where the device already exists, the first step is to disassemble the item into all of its separate parts and identify them with an appropriate functional name. A listing of all these parts is what normally appears in the bill of materials. The one item

17

## Design for Assembly

not considered in this example is a small amount of grease applied to the toggle to insure its smooth motion.

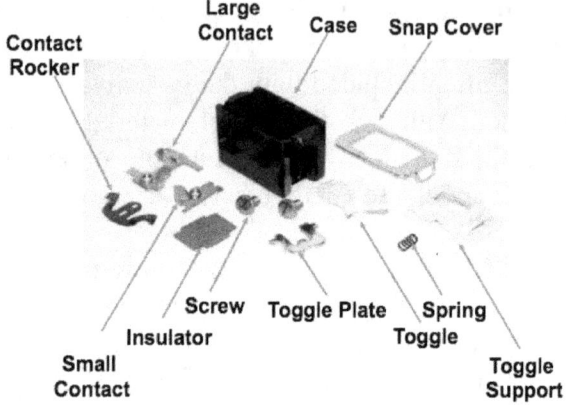

*Figure 1-25: Dis-assembled Toggle Switch*

Presented in Figure 1-26 is a list of all the parts in this simple toggle snap switch. This is in effect the bill of materials. Each part is named with a descriptor that helps identify its function. It is of interest that even this simple device contains twelve separate and distinct parts, excluding the grease.

| Part Number | Number | Name |
|---|---|---|
| 1. | 1 | Case |
| 2. | 1 | Small Contact |
| 3. | 1 | Large Contact |
| 4. | 1 | Insulator |
| 5. | 2 | Wire Screws |
| 6. | 1 | Toggle Spring |
| 7. | 1 | Contact Rocker |
| 8. | 1 | Toggle plate |
| 9. | 1 | Toggle |
| 10. | 1 | Toggle Support |
| 11. | <u>1</u> | Snap Cover |
| Total | 12 | |

*Figure 1-26: Parts List*

Design for Assembly

The multiplicity of parts in any assembly is further justification for establishing a part sequence diagram to indicate the progressive sequence of assembly operations required to produce a final product.

The switch will now be reassembled to provide the sequence of operations needed to create the part sequence diagram. The switch case provides a convenient base component into which the other parts are all inserted to form the final product. In Step 1 the large contactor is inserted vertically down into the left end of the case.

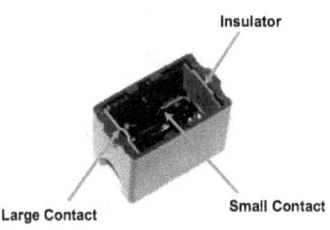

*Figure 1-27: Switch Case*

This is followed by the vertical insertion of the small contact at the right end along with the insulation.

Figure 1-28 depicts the beginning of the part sequence diagram. Step 1 as described is represented by the activity stage Step 1 in which the large contact, small contact and insulator are inserted into the case as the base component. This allows activity Step 2 to be undertaken next which is the insertion of the conducting wire capture screws into the two contactor plates that have been previously drilled and tapped.

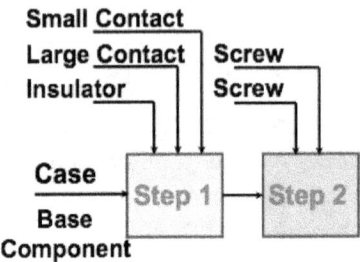

*Figure 1-28: Part Sequence Step 1 & 2*

In preparation for activity Step 3 a subassembly consisting of the contact rocker and the toggle plate must first be composed. When this subassembly is inserted into the case it will permit the toggle plate to rotate about the toggle plate axis that in turn makes or breaks contact by the attached contact rocker arms with the large and small contact plates.

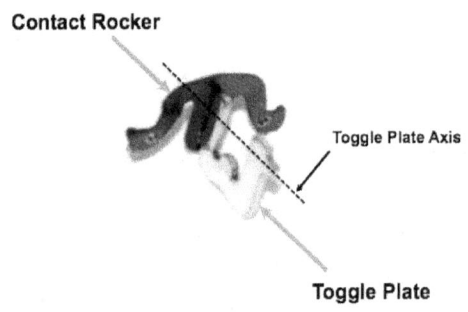

*Figure 1-29: Subassembly for Step 3*

Design for Assembly

In activity Step 3 the toggle plate rocker arm sub-assembly is inserted into the base component with the toggle plate axle tabs placed in slots on either side of the case top. This permits the toggle plate to rotate so that the rocker arm makes and breaks contact with contactor plates at the two ends of the case. The final activity in Step 3 is to place one end of the coil spring on the tab at the bottom of the toggle plate. Figure 1-30 shows Step 3 completed.

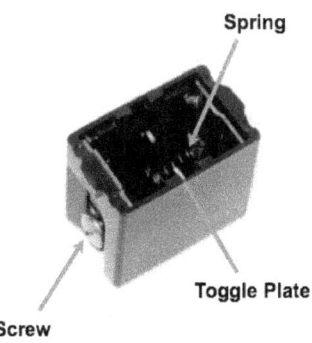

*Figure 1-30: Step 3*

Figure 1–31 shows activity Step 3 added to the developing part sequence diagram. It includes the creation of the toggle plate rocker arm sub assembly prior to its insertion into the base component. Also depicted is the insertion of one end of the helical spring onto the tab of the toggle plate.

Design for Assembly

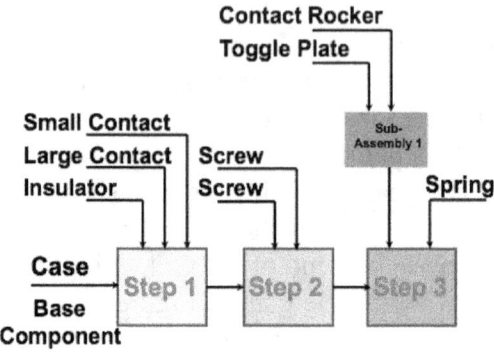

*Figure 1-31: Part Sequence Step 3*

Composing the toggle to the toggle support base now creates a second subassembly in Figure 1-32.

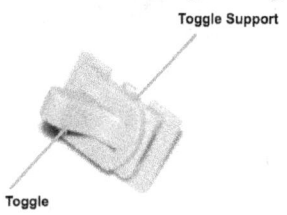

*Figure 1-32: Subassembly for Step 4*

Not visible here is the tab on the bottom of the toggle that must be placed into the free end of the helical spring already in the assembly.

In activity Step 4 subassembly 2 is first composed before it is inserted into the base component. Its insertion is a difficult step since the tab on the bottom

Design for Assembly

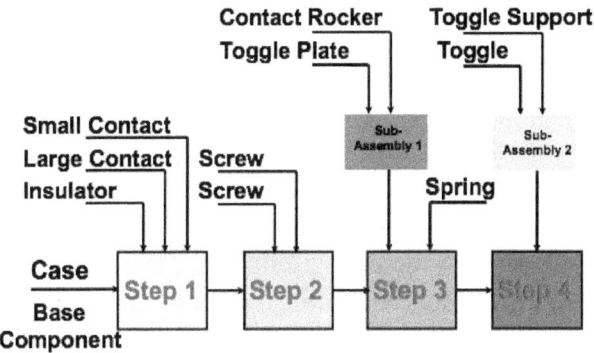

*Figure 1-33: Part Sequence Step 4*

of the toggle must be fitted blindly into the open end of the helical spring that is placed in compression as the toggle support is held against the top of the case.

To provide final composition of the switch the snap cover must be placed over subassembly 2 and snapped into place on the side tabs of the case. This last action holds the entire device together.

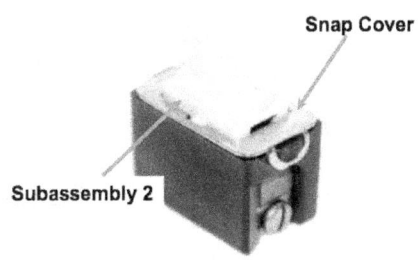

*Figure 1-34: Final Composition*

23

# Design for Assembly

As the toggle is moved back and forth the compressed helical spring causes the toggle plate and rocker arm to snap to open and closed operational positions.

The part sequence diagram is completed with the addition of activity Step 5 that inserts and composes the snap cover with the case producing the finished switch. It is observed that even in this simple device consisting of only 5 activity steps and two subassembly creations that the order of composition must be correct to insure an operational final device.

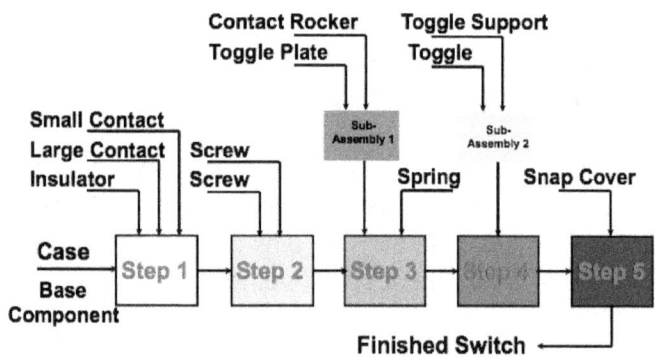

*Figure 1-35: Part Sequence Step 5*

Creating this type of assembly part sequence diagram is particularly helpful in the application of design for assembly principles in product redesign and improvement that will increase functionality and lower production costs.

# Chapter 2 – Product Development Guidelines

Chapter 2 deals with the principles that underlie the application of DFA in the design of new products or the improvement of existing devices.

The DFA principles to be presented are best subdivided into two categories as they relate to the purpose of the principles. The first category deals with product design principles and is directed toward the ease of assembly of the entire devise. The second category deals with the design of parts to facilitate their feeding, orientation and insertion. This chapter covers product design principles.

- Product Design – **for ease of assembly**
- Design of Parts – **for feeding and orienting**

*Figure 2-1: DFA Principles*

## Principles of Product Design

There are no physical laws that govern DFA like those that exist for engineering mechanics. What has been learned about DFA and its application has come from experience and observation in the process of conducting good engineering design. The technical literature contains many references on the use of DFA. Virtually all of them contain lists of guidelines or principles for its application. In most instances these lists contain many similar concepts and are quite repetitive although the number and classification of guidelines differ. This presentation is a distillation and amalgamation of these lists into seven simple and concise principles that reasonably cover what is known and referenced about the subject. These principles apply specifically to considerations that affect the ease with which a product can be assembled. Simply stated they are: the assembly should be built on a suitable base, components and parts should be layered and stacked in sequence, parts and subsystems should be modular if possible, errors and uncertainties need to be accommodated, simplifying and reducing fasteners is always a prime goal, flexible items should be limited and finally part count should be minimized.

1. Build assembly on a suitable base
2. Layer and stack assembly
3. Use modular and multi-functional parts
4. Accommodate for errors and uncertainties
5. Simplify and reduce number of fasteners
6. Limit flexible items
7. Minimize number of parts

*Figure 2-2: Principles of Product Design*

## Importance of Base

Providing a suitable base promotes ease of assembly in that it can serve as a base component into which other parts are inserted as the assembly comes together. Characteristics of a good base component are that it be oriented in a horizontal plane and can provide stable positioning and support for part insertion and composition. It should also establish part positions with guide aids like pins, slots, recesses, etc.

# Design for Assembly

Orient in a horizontal plane

Provide stable position and support

Establish part positions with guide aids
- pins
- recesses
- inserts . . . etc.

*Figure 2-3: Advantages of Providing a Base*

In this way special fixtures required to assist in the assembly are eliminated. For example a bicycle frame serves as the base component for assembling the complete device. However a special fixture is required to hold the frame as parts are added. In assembling a smart phone the back case of the device becomes a suitable device that possesses the orientation and support characteristic listed without requiring any additional fixtures.

The two examples shown here illustrate suitable but different base components for two 120-volt residential electric wall receptacles.

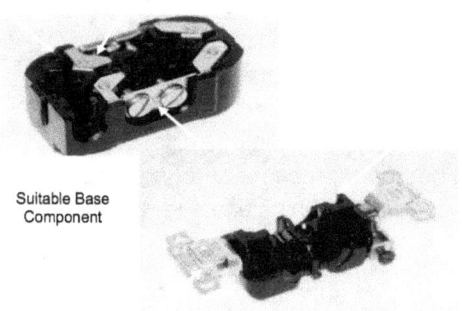

Suitable Base Component

*Figure 2-4: Example of Suitable Bases*

27

On the bottom right the utility box attachment strap with the back base of the receptacle serve as the base component. The electrical contact components can now be inserted into the back base and then the case top added. In the upper left the front or top cover of the receptacle serves as the base component into which the electrical contact components have already been inserted. The back of the case can now be added and the utility box strap inserted and composed over that. Note in this example how the front cover and back case were molded to provide exact location positions for the electrical contact components.

## Layered Assembly

Layering the assembly is emphasized in DFA employment in order to simplify insertion and provide for appropriate sequencing. With the base component ideally oriented horizontally parts should be stacked vertically with insertion taking place from above. Where possible subassemblies should be created that will promote vertical insertion. Pursue designs that provide for easy insertion and disassembly.

> Place base on assembly fixture first
> Stack parts vertically
> Assemble from above
> Create subassemblies for convenience
> Insure easy insertion and disassembly
> Do not rotate, reorient or regrip parts
> Orient and apply fasteners vertically
> Let *gravity* help

*Figure 2-5: Layering*

## Design for Assembly

Avoid needing to rotate, reorient or re-grip parts during the insertion process. Ideally fasteners should be oriented and applied vertically from above. The whole idea is to let gravity help make assembly easier.

Shown at the top left in Figure 2-6 is the wall receptacle of the previous example that uses the top cover as the base component. The assembly of the device takes place with all components inserted vertically down satisfying a number of principles of layered assembly. Note that the electrical contacts and the utility box support strap are subassemblies that have been separately prepared and made available for final assembly. At the lower right is the receptacle example that uses the support strap and case bottom as the base component. Again the design illustrates design for layered assembly with the strap and electrical contact elements brought to final assembly as subassemblies.

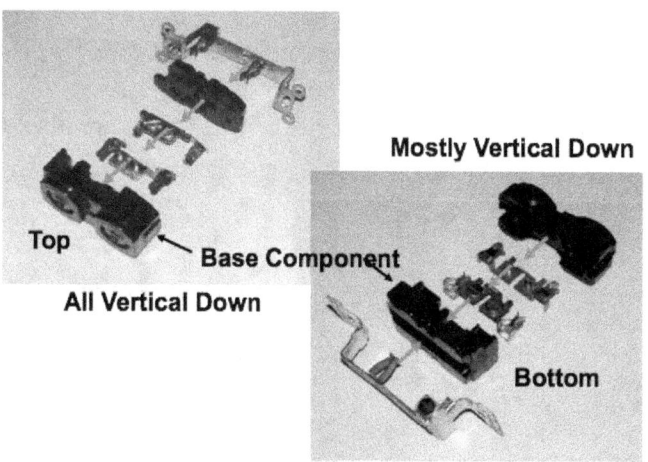

*Figure 2-6: Example of Layering*

Design for Assembly

## Modular Components

The modular component DFA guideline emphasizes the grouping of parts into modular subassemblies that can be used functionally for similar models within the same family of products. The electrical contact subassemblies of the previous examples are simple illustrations of modular units. These subassemblies make final product assembly easier, reducing assembly time and cost. They also reduce assembly complexity and improve final product quality.

> Group components into functional modules for similar products in the same family
> Subassemble for use in final product
> Reduces final assembly time
> Reduces final assembly complexity
> Increases final product quality

*Figure 2-7: Modular Components Summary*

Two examples of very complex modular units is I illustrated in Figure 2-8. Both the starter motor and alternator are products in their own right but are also

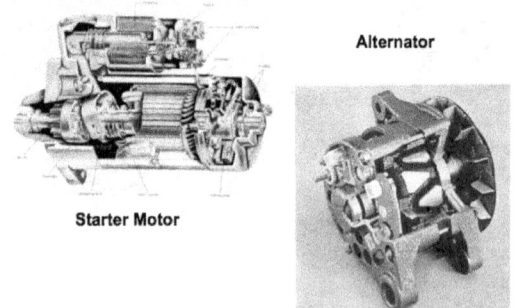

*Figure 2-8  Example Modular Components*

modular units for use with different models of engines for installation in cars built by different automotive and truck manufactures.

## Accommodation for Errors

The accommodation for errors DFA guideline deals with assembly insertion problems created by part tolerance, uncertainty in dimensional stability, mechanical wear and alignment issues. Many of these can be overcome with design considerations that account for individual part and stacking tolerance issues in initial part dimension specifications. Insertion and composition is always improved and made easier by attention given to designing parts to be self-aligning. This is achieved by taking advantage of the benefits of chamfers, ramps or lead-ins along with guide pins, notched slots, cut outs, tabs and dimples. The next several pages illustrate this with some simple examples.

Overcome uncertainty in dimensional
   stability, part tolerance and mechanical wear

Design parts to be self aligning

Use fitting aids
   - chamfers, ramps, lead ins . . .

Orientation Guides
   - guide pins, d-shaped holes, notched slots
   cut outs, tabs, perforations, dimples . . .

*Figure 2-9: Accommodation for Errors*

The so-called "pin in hole" problem or some close variation to it is one of the most commonly encountered insertion activities

Design for Assembly

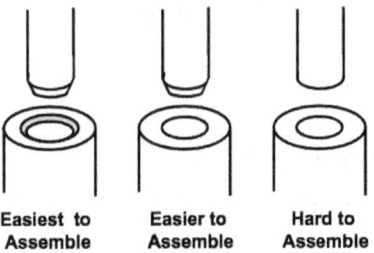

*Figure 2-10: Pin in Hole Assembly*

For close tolerance requirements the simple use of chamfers as illustrated in Figure 2-10 will significantly make the insertion activity easier.

In the example shown in Figure 2-11 inserting the collar vertically down onto the washer and pin and compressing the spring on the left requires multiple initial alignment requirements. These can be eliminated as shown on the right simply by extending the length of the pin so that the washer and collar don't have to be aligned together as the collar is inserted to produce the required spring compression. Chamfering the pin also aids the assembly process.

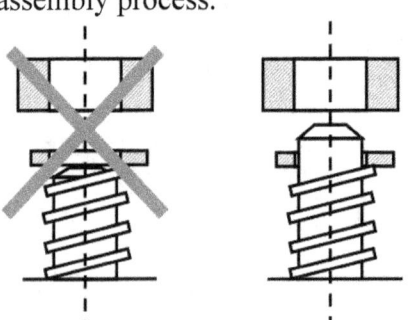

*Figure 2-11: Alignment Example 1*

Design for Assembly

On the left in Figure 2-12 the difficulty in achieving insertion with a pin that requires simultaneous alignment a two different hole locations is obvious. This might be further complicated if the lower hole were blind to the installer. On the right the problem is easily solved by having the insertion at the upper hole take place first automatically providing alignment at the lower hole by a simple dimensional change in the length of the lower portion of the pin.

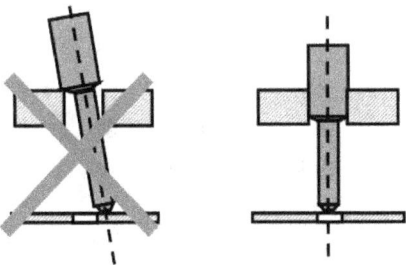

*Figure 2-12: Alignment Example 2*

## Fastener Reduction

The number and variety of common and specialized fasteners available to designers is almost limitless. This provides for flexibility and creativity in how composition can be accomplished in product design. This availability and the tendency to overdesign the quality of the composition frequently leads to fasteners representing a significant percentage of the total number of parts in many products. The goal of the DFA guideline for simplifying and reducing fasteners is

- Eliminate screws, bolts, clips, rivets, etc. where possible
- Use snaps, tabs, interlocks, slots & press fits
- If screws necessary, consider self tapping
- Orient for vertical insertion
- Provide room for automatic tools
- Keep field maintenance tooling in mind
- Adhesives or welding simplify design and assembly but complicate repair

*Figure 2-13: Fastener reduction summary*

directed toward taking advantage of the benefits of new types of fasteners while reducing total fastener part count. No more should be used than are absolutely necessary to insure quality of composition. Some of the more important suggestions that can help achieve this goal are listed in Figure 2-13. Eliminate separate fasteners where possible particularly when composition is over designed and replace with integral devices like snaps, tabs, slots, press fits, etc. If screws are necessary consider self-tapping and orient for vertical insertion. Keep in mind that sufficient room is required for automatic tooling and tooling for field maintenance can be limited. Finally, adhesives and welding can reduce fasteners but may complicate device repair.

Figure 2-14 illustrates how a quality composition is achieved for the sheet metal bracket by use of a simple tab and slot and a single self-tapping sheet metal screw. By contrast a classic composition technique might specify four nut, bolt and washer combinations at the four corners of the bracket base totaling twelve parts to the two shown in the more creative solution.

Design for Assembly

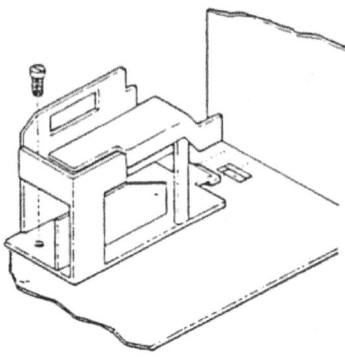

*Figure 2-14: Fastener Reduction Example*

Depicted in Figure 2-15 is the geometry of a generic snap fastener that simply engages the snap hook with the receiver detent when insertion takes place. As illustrated the receiver must provide sufficient space for the required deformation of the snap and an adequate lead in for the snap hook.

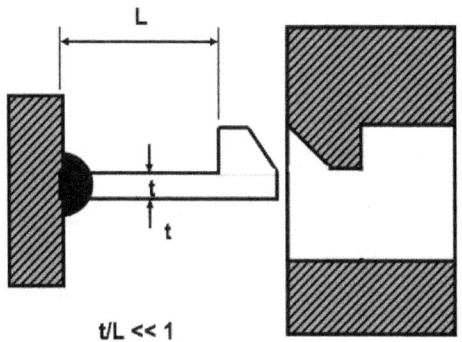

*Figure 2-15: Snap Fastener*

With the increasing use of plastics as a material of choice in product development the integral snap fastener

35

has become the number one replacement for individual fasteners. It provides a quality composition without adding to the total part count. This innovative variation of the snap fastener concept in Figure 2-16 is unique in that the hinge point is at the front rather than the rear as in the generic depiction. This permits easy multiple insertions and extractions which is a functional requirement. Extraction is achieved by use of the rear extension of the snap permitting easy compression with thumb and finger to disengage the snap hook from the receiver detent.

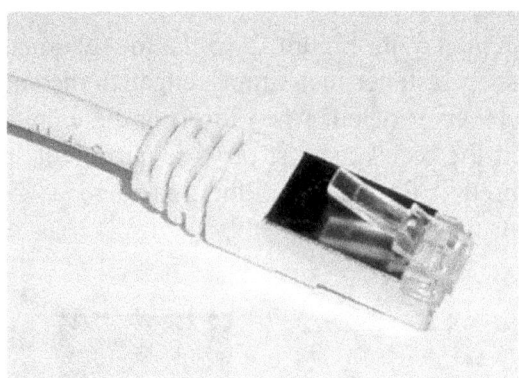

*Figure 2-16: LAN Snap Application*

Shown in Figure 2-17 is another variation of the snap fastener concept used in composing cover panels to structural frames like door panels in an automobile. A simple circular hole in the structure serves as the receiver. The insertion and extraction ramp geometry in the design provides all the required flexibility for both attachment and withdrawal of the fastener.

Design for Assembly

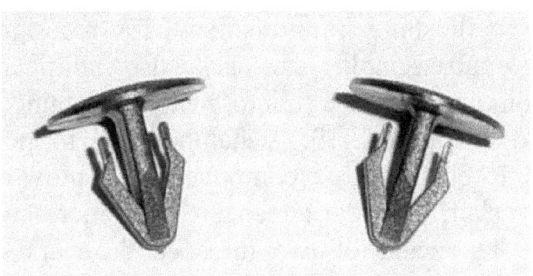

*Figure 2-17: Door Panels in Automobiles*

## Minimizing Parts

The consequence of the minimizing parts DFA guideline when applied to product design is reduction of costs in multiple areas. With fewer parts all handing functions in the assembly process are reduced. This results in simplifying assembly and reducing assembly time. Fewer parts means reduced manufacturing requirements. The fewer the parts the simpler the product. This increases quality, reliability and maintainability. All of these effects will reduce the overall cost of the final product. The challenge in achieving this end rests with the creative and innovative capabilities of the designer whether being applied to new product development or improvement of existing devices.

- Simplifies assembly
- Reduces manufacturing costs
- Reduces assembly time
- Increases quality
- Improves reliability and maintainability

*Figure 2-18: Minimization of Parts*

Design for Assembly

To illustrate how creative DFA redesign of an existing subassembly can achieve significant part reduction consider the micro-switch actuating device depicted in Figure 2-19. A stamped and formed sheet metal bracket is the base component. It is provided with a pivot shaft already attached. The micro-switch is attached by means of two threaded fasteners that are captured by a nut plate.

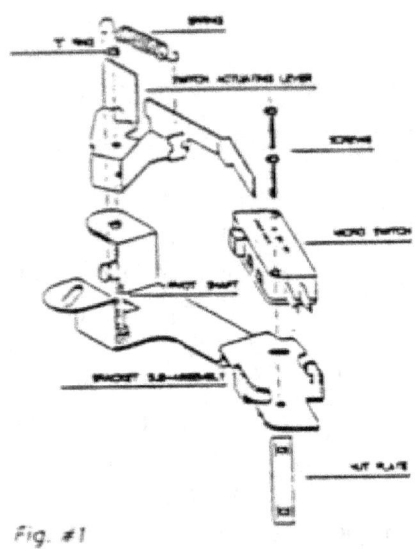

*Figure 2-19: Micro Switching Actuating Device*

The formed sheet metal switch-actuating arm is placed over the pivot shaft and restrained by a circlip. Finally a helical spring that extends from the base component to the switch-actuating arm keeps the switch in a fixed position. This assembly consists of eight separate parts not counting the switch itself.

# Design for Assembly

Shown in Figure 2-20 is an innovative redesign of this subassembly taking advantage of the flexibility and molding capabilities of plastic. This redesign satisfies all the functionality requirements of the original design. The long cantilevered element serves both as the switch-actuating arm and the spring. Fastening of the micro-switch is accomplished by providing two molded stub pins that locate its position and two integral snap fasteners that simply clip over the edge of the micro-switch body. The entire new subassembly not counting the micro switch is one single part.

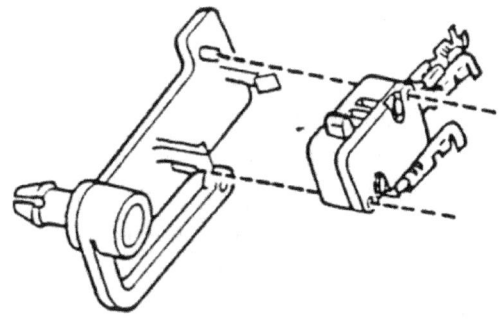

*Figure 2-20: Minimized Parts Using Molding*

Another example of creative redesign that minimizes part count is illustrated in Figure 2-21. On the left is the classic staple remover that when disassembly consists of six separate parts. These include two steel staple claws, a pivot pin, a torsion spring and two plastic actuating handles. Shown on the right is a redesign consisting of only three parts.

# Design for Assembly

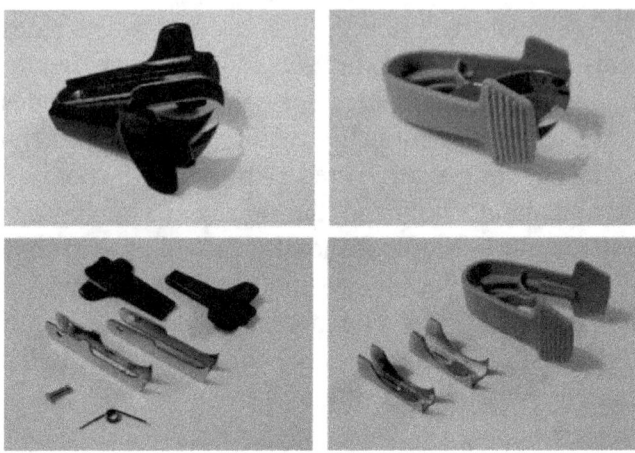

*Figure 2-21: Staple Remover*

The plastic handles, the pivot pin and torsion spring are all functionally included in the single plastic molding that permits the same mechanical action by means of a continuous plastic hinge. The staple removing claws are retained but their attachment is simplified by introducing a slot and hole which provide for location and composition with a dimple and locator ridge molded into the underside of the finger gripper.

Fastening together two steel gears with four threaded fasteners creates the simple two-gear reduction subassembly in the upper diagram of Figure 2-22. This is an assembly of six parts. Assuming that power transfer requirement can be met this assembly can be replaced as shown in the lower figure by use of sintering, plastic molding or casting as a single part that permits the same functional result. A good example of different manufacturing options that results in a creative redesign solution.

Design for Assembly

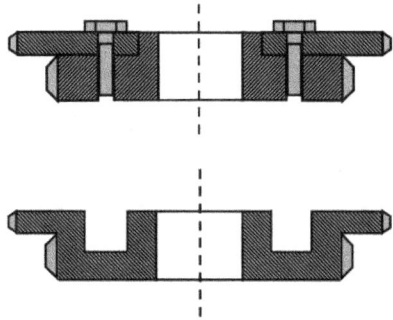

*Figure 2-22: Two gear Reduction Assembly*

Flexible Items

Flexible items introduce composition difficulties whether done manually or by machine. This is particularly true of wires, cables ad belts. Where possible electrical components should be plugged together or mounted on a circuit board. Consideration should be given to replacing belts by gears or friction drives.

> Wires, cables and belts are difficult to handle in automated assembly
>
> Components should be plugged together where possible
>
> Mount electronic components on circuit boards
>
> Use gears and shafts to substitute for belt drives

*Figure 2-23: Flexible Items Design Summary*

Although this may add additional parts and cost, the ease of assembly when composition is mechanized can well be worth it.

Illustrated in Figure 2-24 is the ubiquitous printed circuit board. The development of this technology represents the ultimate application of the DFA principle of eliminating flexible items. It should be taken advantage of whenever possible in the development of a new product or redesign of existing devices.

*Figure 2-24: Circuit Board*

## Part Redundancy

The redundancy criteria presented here can be helpful in determining whether a specific part is really needed in a new design or is redundant in a redesign effort. It stems from the reasons for separate parts covered in Chapter 1 but only makes use of three reasons to provide a starting point.

Three questions are asked of each part. The first question is does the part need to move relative to parts already assembled? The second question is does the part need to be of a different material from parts already assembled that it will be adjacent to?

## Design for Assembly

1. **Does the part move relative to parts already assembled in the function of the final product?**

2. **Does the part need to be made of a different material from parts already assembled for functional (but not aesthetic) reasons?**

3. **Does the part need to be separate from parts already assembled to permit assembly or disassembly?**

"No" answers to all three criteria identifies a part as a potential candidate for elimination

*Figure 2-25: Part Redundancy Criteria*

The final question is does the part need to be separate from parts already assembled to permit assembly or disassembly? If the answer is "NO" to all three questions the part should be considered as a potential candidate for elimination. Keep in mind that this is a simplified criterion that can be helpful but is by no means an absolute determination. Recall that there are reasons why parts should be separate as a consequence of other considerations.

# Design for Assembly

# Chapter 3 – Part Design for Feeding and Insertion

In Chapter 3 the DFA principles and guidelines for the design of parts to facilitate feeding, orientation and insertion are presented as distinct from the DFA product guidelines.

## Four Part Spindle Subassembly

To illustrate the importance of "part" design guidelines consider the assembly of a four-part spindle subassembly. The four parts consist of the spindle body, a drop of Loctite, a gasket and a screw. The assembly tasks include presenting the spindle body, dosing the recessed spindle top with Loctite, adding a gasket, inserting a screw and adjusting the final height of the screw. The individual assembly process steps required to build this subassembly on a mechanized rotary indexing table will now be examined.

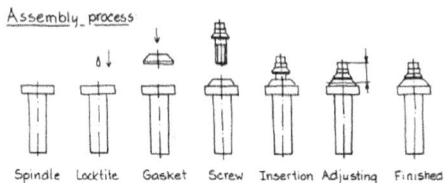

*Figure 3-1: Assembly Process*

## Applying the Loctite

As illustrated in Figure 3-2, dosing the spindle with Loctite takes five separate steps. The spindle bodies stored in bulk must first be fed by a vibratory feeder through a passive orienting device. They are then delivered by a horizontal transporting rail to where one is allocated into a holding position on the rotary table. The table indexes to the next station where a drop of Loctite is deposited in the recessed top of the spindle.

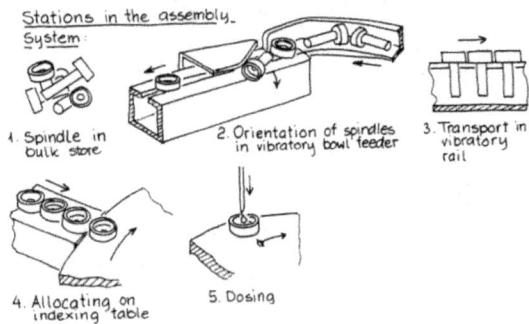

*Figure 3-2: Applying the Loctite*

## Gasket Addition

Coming from bulk storage the gaskets are feed by a second vibratory feeder through another passive orienting device.

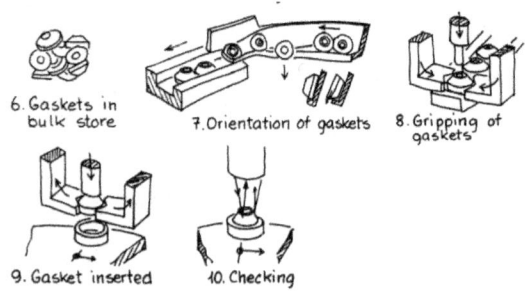

*Figure 3-3: Gasket Addition*

# Design for Assembly

Gasket allocation is provided by means of side grippers and a vertical locating pin that inserts the gasket when properly positioned over the spindle top. The final step is an optical checking of the partial subassembly to determine that a gasket is present. The gasket addition takes an additional five steps.

## Screw Insertion & Adjustment

To finish the subassembly seven more steps are required. The screws again come from bulk storage. A third vibratory feeder passes the screws through another passive orienting device to a transporting rail to their allocation position. At this point they are griped and then inserted through the gasket and screwed into the spindle body. At the final assembly station an adjustment gauge checks the screw height and makes any corrections necessary.

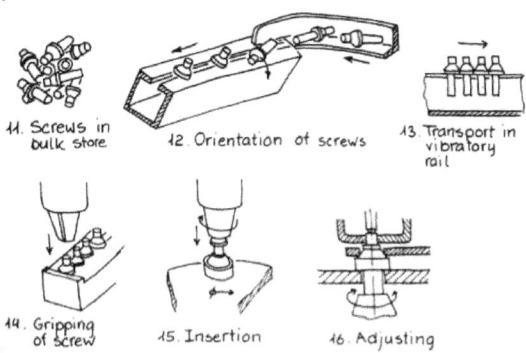

*Figure 3-4: Screw Insertion & Adjustment*

## Assembly Sequence Diagram

The entire process of creating this spindle subassembly is represented here by an assembly sequence flow diagram covering all 17 required steps. The generic assembly activities of presenting a spindle, dosing it with Loctite and inserting and adjusting the screw are shown on the right by the heavy shaded blocks. All of the remaining lighter blocks represent the variety of activities and actions that each of the separate parts undergo in the process of producing the final product. This illustrates why DFA principles and guidelines for part design to facilitate feeding and orientation is as important as the DFA product guidelines. This is just as true for manual assembly as it is for automated assembly.

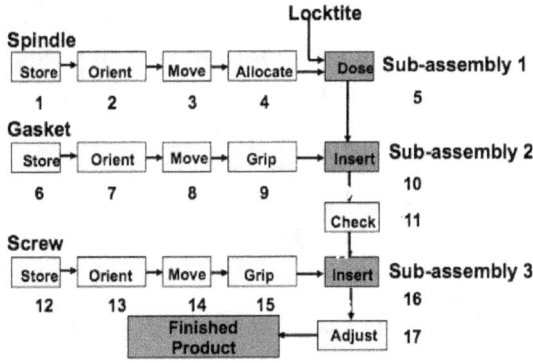

*Figure 3-5: Assembly Sequence Diagram*

## Part Design Guidelines

The purpose and prime objective of the DFA part design guidelines is to facilitate feeding and orienting since these activities play such a large part in the assembly process. The most important considerations can be summarized in three categories. The first is to avoid geometric shapes or configurations that can lead to tangling, overlapping and nesting of parts. The second is

## Design for Assembly

to be aware of how gravity can both create difficulties and assist in final part orientation. The third category deals with how emphasizing symmetry or taking advantage of asymmetry in external part geometry can be used to promote orientation.

> **Prime Objective is to facilitate feeding and orientation**
> Avoid or Eliminate
> **Tangling, overlapping, and nesting**
> Gravity
> **Consider center of gravity location**
> Symmetry
> **Emphasize or use asymmetry**

*Figure 3-6: Part Design Guidelines*

## Tangling, Overlapping and Nesting

Listed in Figure 3-7 are six actions that can be taken to reduce tangling, overlapping and nesting.

> **Eliminate protrusions that can enter holes slots or open penetrations**
>
> **Close coil spring ends or produce at site**
>
> **Provide thicker contact surfaces for thin flat elements that might shingle**
>
> **Reduce opening size in retaining rings, lock washers, etc.**
>
> **Increase angles on flat element that might overlap**
>
> **Use ribs to stop nesting**

*Figure 3-7: Tangling, Overlapping & Nesting*

They include eliminating protrusions in one part that can enter into holes, slots or open penetrations in adjacent parts. Closing coil spring ends, reducing coil spacing or

## Design for Assembly

producing the part at the site of the assembly can help prevent tangling. Providing thicker surfaces for thin flat elements that might shingle over one another. Keeping opening sizes in retaining rings, lock washers and like element to a minimum. Increasing angles or edge geometry on flat element that might overlap. Also using spacers or ribs to keep parts from nesting. All of these actions and conditions will be demonstrated in the following examples.

### Protrusion Example 1

Shown in Figure 3-8 are two mounting straps from a single pole residential wall switch. The designer has made the dimension cap D larger than the dimension little d so that the end of the two strap will not get entangled when being fed from some bulk source.

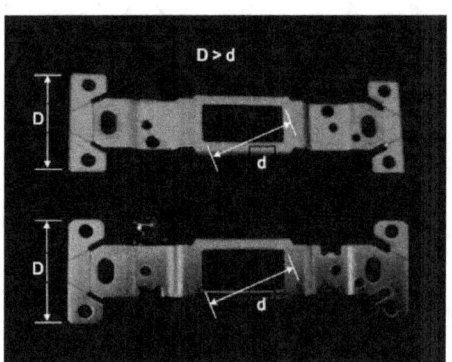

*Figure 3-8: Protrusion Example 1*

Unfortunately tangling can and will still occur because of the relief slot that permits the end tab to pass through the center rectangular hole. Like Murphy's Law if tangling can happen it will even though it doesn't seem possible (Figure 3-9)

Design for Assembly

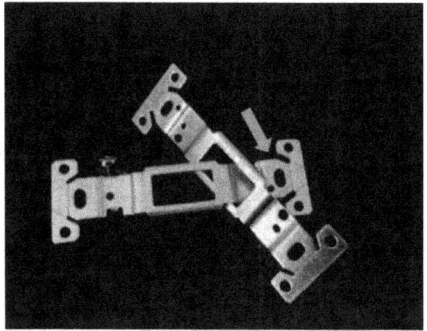

*Figure 3-9: Example 1 Tangling*

## Protrusion Example 2

The solution to a potential tangling problem is illustrated here. The part on the left with the slot can tangle in bulk since the width of the lower tab is smaller than the length of the slot and the tab thickness is less than the width of the slot. A correction to this problem might simply be to add a dimple as shown, which would keep this from happening.

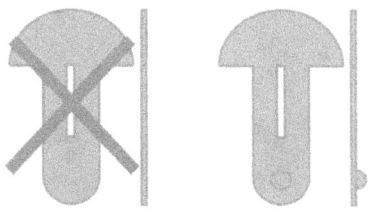

*Figure 3-10: Protrusion Example 2*

It could be introduced as part of the blanking process that most likely produced the part. Another consideration would have to be that the dimple did not affect the function of the original part.

51

## Tangling Example 1

Coil springs are the hardest parts to feed without tangling whether by machine or manually. The two geometric characteristics that create problems are hook ends and coil spacing. Both are illustrated here with a number of examples in Figure 3-11. Note the tension spring at the center right that solves these problems with closed spring ends and no coil spacing. This is fine for tension springs but obviously compression springs must be provided with finite coil spacing to function properly. Creative design is required to eliminate this problem. It may be that some other type of spring should be used like the plastic hinge in the stapler redesign.

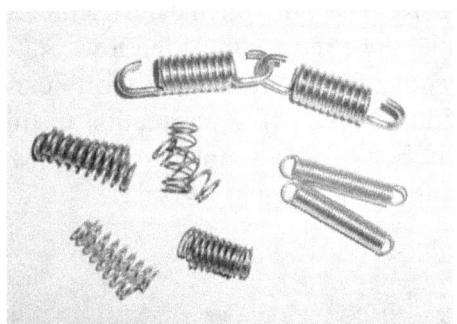

*Figure 3-11: Coil Springs*

## Tangling Example 2

Another classic tangling problem are split rings and washers. If the gap in the circumference is large enough tangling will occur. This is particular true on internal circlips where sufficient gap space must be provided for ease of insertion. As shown in Figure 3-12 this is not a problem with external circlips. This is just one example where the designer's creativity and

Design for Assembly

innovation is called into play to provide composition capability without introducing part-feeding problems.

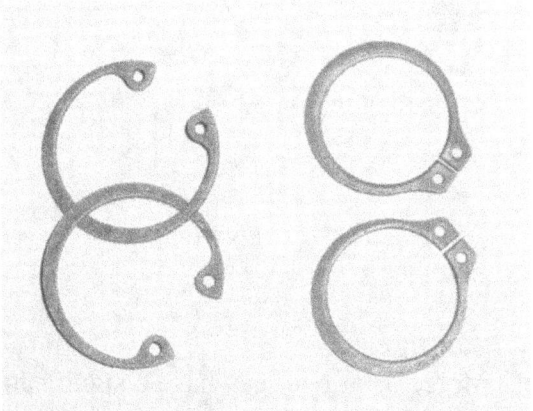

*Figure 3-12: Circlips*

## Overlapping Example

Shown in Figure 3-13 is an example of how a rivet like fastener can jam in gravity fed delivery track due to the shape of the head. Possible solutions to the problem are shown by simple modifications of the head geometry at the upper left and lower right. The key to the solution of potential feeding problems is the recognition of the problem beforehand. This is not always immediately obvious but deserves adequate attention at the part design phase of product development or redesign.

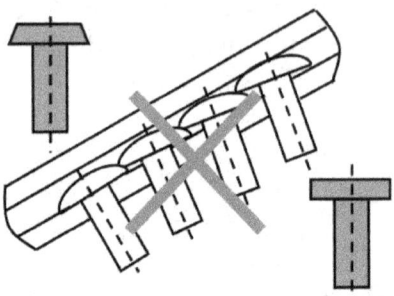

*Figure 3-13: Overlapping*

## Gravity Example 1

The force of gravity can have a significant effect on part orientation. Passing through the center of gravity of the part it always acts vertically down. This can be both beneficial and/or detrimental depending on what the final position is desired for the part. In the example shown here on the right that is to be oriented vertically up the width of the base may not provide sufficient restraint to the part from being tipped over. Easily a problem in the vibratory shaking of a bowl feeder since the center of gravity is located in the upper portion of the part.

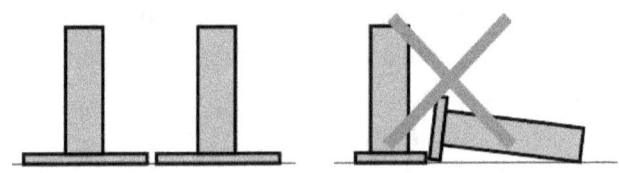

*Figure 3-14: Gravity Example 1*

Design for Assembly

One way to remedy this problem would be to widen the base as shown on the left provided this did not adversely impact the functionality of the part.

## Gravity Example 2

This example shows how gravity is used to orient a headed bolt as it is being fed through the end of an orienting device n a vibratory bowl. The center of gravity located along the shank of the bolt automatically causes the bolt to orient vertically down with the head on top due to the action of gravity.

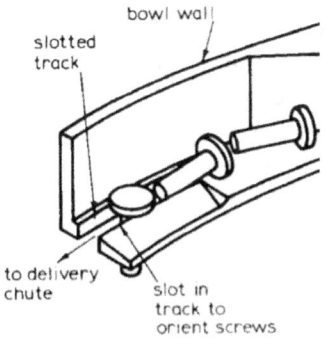

*Figure 3-15: Gravity Example 2*

## Overlooked Example 1

Very often unique and creative part feeding and orienting examples present themselves unexpectedly. The packaging of ordinary staples is a good example. By gluing the staples together into a strip and designing a stapler to accept this package a difficult orienting and feeding problem is uniquely solved.

Design for Assembly

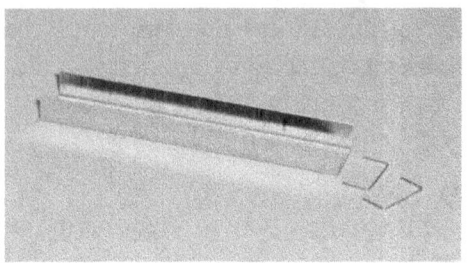

*Figure 3-16: Staples*

## Overlooked Example 2

The packaging and dispensing of self-adhesive postage stamps on a paper roll for individual manual separation and use is another creative example. Placing miniature electronic components on tape to be fed into an automated circuit board insertion machine makes use of the same principle.

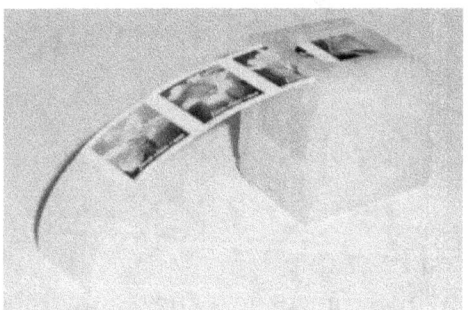

*Figure 3-17: Postage Stamps*

## Nesting and Clinging Example

Depending on material properties, surface conditions and geometry part separation and feeding can present problems due to clinging when parts are nested.

Design for Assembly

Although not strictly an assembly problem a common example of this occurring is the sticking together of a stack of Styrofoam cups and the subsequent problem of their separation. Another example would be layers or stacks of parts where the surfaces tended to cling together.

*Figure 3-18: Styrofoam cups*

## Symmetry

Since most automated part orientation depends on the passive sensing of eternal part geometry symmetry can play a very important part in the feeding and orienting of small parts. Some of the more important benefits, consequences and applications are listed here.

**Benefits greatest about axis of insertion**

**Parts can be fed more easily**

**Reduces handling to orient parts**

**Simplifies picking and holding**

**Can reduce manufacturing costs**

*Figure 3-19: Symmetry*

# Design for Assembly

The greatest benefits of symmetry are derived when it can be applied about the axis of insertion of the part. The picking, handling and holding of such parts is simplified whether done manually or automatically, parts are generally easier to insert the more symmetrical their geometry and manufacturing costs can also be reduced as a consequence of symmetry in the part geometry.

## Symmetry Example 1

The two ends being different on the stud on the right will increase the cost of manufacture and will require special handling and orientation if one end of the pin is always inserted first. For functional reasons this may be necessary. However, if the overall design were modified so that each end of the pin could be the same as on the left then both manufacturing and assembly would be improved. No special end for end orientation would be required for either operation.

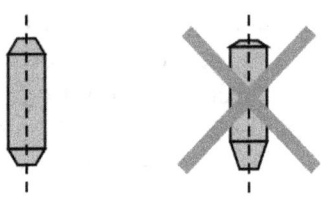

*Figure 3-20: Symmetry Example 1*

## Symmetry Example 2

Shown in Figure 3-21 are three treaded studs removed from a small vintage international four-cylinder engine block. It is observed that the treaded lengths on each stud are different from one end to the other. One must wonder why this was done since making the treaded lengths equal would have made end for end orientation easier for the assembler.

*Figure 3-21: Symmetry Example 2*

## Symmetry Example 3

This part as manufactured on the right possesses no symmetry about either axis. It will require special handling about both axes to provide the proper orientation for insertion and composition.

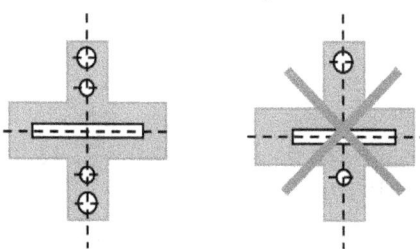

*Figure 3-22: Symmetry Example 3*

A possible redesign change that would make handling and orientation much simpler would be to force symmetry on the part even though all new aspects might not be functional in the overall design. This is illustrated on the left where symmetry now exist with respect to both axes simply by adding two extra holes and changing exterior geometry of the horizontal portion. If this was a stamped part manufacturing would not be affected but assembly would be enhanced.

## Asymmetry

In some instances when symmetry can't be achieved or does not provide the desired result it is worthwhile emphasizing asymmetry in the external geometry. This is because passive orienting devices used in automated assembly generally work on the principle of the device sensing some external asymmetric feature of the part. To take advantage of asymmetry it is well to exaggerate small feature of asymmetric design. Sometime it is appropriate e to deliberately add features that are asymmetric. Another technique used to achieve asymmetry is to remove material from the external geometry to produce a significant feature.

**Exaggerate small features of asymmetric design**

**Deliberately add features which are asymmetric**

**Remove material to produce asymmetry or significant external feature**

*Figure 3-23: Asymmetry*

Design for Assembly

## Asymmetry Example 1

The washer on the left with a small hole on the vertical axis would be impossible to orient with an automatic passive device as there is nothing to establish a specific orientation from the circular exterior geometry. Even manual orientation requires visual recognition to establish the specific small hole positioning.

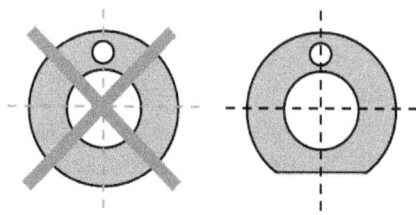

*Figure 3-24: Asymmetry Example 1*

Removing some material from the bottom of the washer on the right perpendicular to the axis through the small hole now provides an external asymmetric feature that can be detected by a passive orienting device or manually without requiring a visual cue.

## Asymmetry Example 2

The small power tool motor fan blade in Figure 3-25 takes advantage of the asymmetric design of the front and backsides to achieve ease of orientation as it passes through a passive orientation device on an automated assembly line. An inclined trough simply eliminates one orientation while allowing the other to pass on.

Design for Assembly

*Figure 3-25: Asymmetry Example 2*

## Asymmetry Example 3

Although the part on the left in Figure 3-26 has an asymmetric machined bevel on one end this is not a feature that will be easily recognized by a passive orienting device. To resolve this issue two different approaches can be applied. In the top graphic on the right a machined bevel is also included on the left end. This makes the part end for end symmetric, which eliminates end for end rotation for orientation. In the bottom right graphic the part is provided with a significant asymmetric feature by machining away part of the cylindrical body. Gravity could be used to provide easy vertical orientation if desired. Either change would make orientation easier.

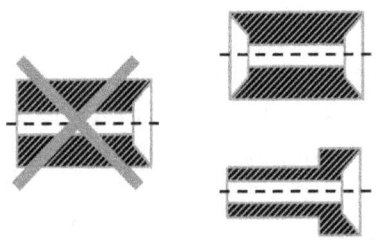

Figure 3.26: Asymmetry Example

# Chapter 4 – Product Assembly Improvement Process

Chapter 4 deals with the process and application of DFA principles to improve the ease of assembly of existing products.

## Product Redesign Process

The principles of DFA should be applied in the earliest stages of product development for maximum effectiveness. However, to gage the potential impact of DFA it is more instructive to consider how ease of assembly can be improved by examining DFA changes in existing designs. A four-step process will be employed to investigate in detail the application of DFA after the fact in several existing designs. The first step in the process is to disassemble the device into all its separate parts, create a parts list and an assembly sequence diagram.

> Separate assembly into component parts
>
> Perform DFA systems analysis
>
> Develop design change candidates
>
> Evaluate redesign for improvements

*Figure 4-1: Product Redesign Process*

The second step is to perform a DFA systems analysis. Consider how the DFA principles apply to this device. The third step is to develop design change candidates that will improve ease of assembly. The

Design for Assembly

fourth and final step is to come up with a measure to evaluate the improvement.

## Damper Valve

Shown in Figure 4-2 is a disassembled damper valve subassembly of the type that would be used to control flow rate in a circular air duct. It consists of eight total parts as listed on the left. These include the damper valve plate that provides the area restriction, the axle housing that permits the damper to rotate and two machine screws along with two washers and two nuts to fasten the axle housing to the damper plate.

**Parts List**
  2 machine screws
  1 axle housing
  1 damper plate
  2 washers
  2 nuts

  8 parts total

*Figure 4-2: Original Damper Valve*

The assembly of the damper valve begins with the damper plate as the base component. The first step is to place the axle housing on the plate and line up the holes in both parts. The next step is to insert the two machine screws through the holes. In step 3 the washers and nuts are added and tightened to complete the final product. This sequencing is depicted in Figure 4-3.

Design for Assembly

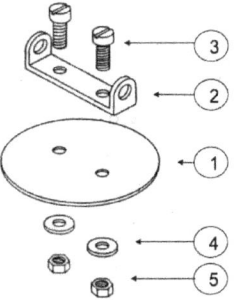

*Figure 4-3: Assembly Damper Valve*

## Part Sequence Diagram

Shown graphically in Figure 4-4 is the part sequence diagram of the assembly process as described. The next task is to apply the DFA guidelines to determine what changes might be made to improve and simplify assembly.

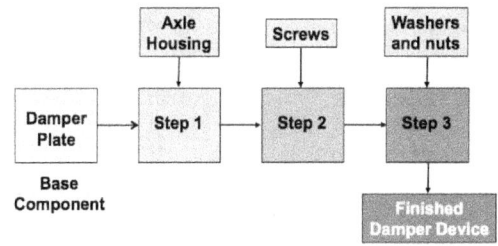

*Figure 4-4: Part Sequence Diagram*

## DFA Systems Analysis

The first design issue that becomes immediately evident is the fastener components number 6 of the 8 total parts or 75%. Applying the DFA guideline to eliminate fasteners and reduce

65

parts represents a fruitful area for a design change. The remaining two parts, the damper plate and axle housing are critical to the function of the device and cannot be candidates for elimination. It is observed that the two holes in the damper plate are interior asymmetric features that will require visual orientation whether assembly is manual or automated. A related feature of the damper plate is that no advantage has been taken of its circular exterior symmetry. Finally composition of the axle housing with the fasteners will be difficult. If gravity is used to hold the machine screws vertical with the threaded shank down the washer and nuts have to be attached from the underside of the damper plate. This action will be awkward and difficult whether attempted manually or automatically.

- Fastener components account for 6 of 8 total parts (75%)
- Neither axle housing or damper plate are candidates for elimination (functional requirements are distinct)
- Two holes in damper plate are asymmetric interior features (orientation will be difficult)
- No advantage taken of circular symmetry of damper plate
- Composition with fasteners will be difficult

*Figure 4-5: DFA Systems Analysis Summary*

## Design Change Candidates

To reduce the number of fasteners two studs could be formed on the axle housing, inserted through the damper holes and be staked to provide a robust composition. This would eliminate all fasteners accounting for six parts. It seems reasonable to go one step further and put a single hole in the center of the damper plate and have only one stud on the axle housing. This would eliminate any special orientation requirements for the damper plate. Finally inverting the order of assembly by having the plate lowered over the axle-housing stub would permit easier staking. With these changes not only is assembly simplified but also costs would be reduced and assembly would lend itself easily to automation.

> Eliminate machine screws, washers and nuts; replace with studs and stake assembly (eliminates 6 parts)
>
> Use only one stud (simplifies orientation since damper plate is circular)
>
> Place stud on axle housing
>
> Invert order of assembly to permit easy staking

*Figure 4-6: Design Change Candidates Summary*

## Redesigned Damper Valve

Shown in Figure 4-7 is the damper device incorporating the proposed changes. The 8 part original design has been reduced to just two functional parts through application of DFA principles. One quantitative measure of the improvement in the original design is a part reduction of 75%. There will also be significant

Design for Assembly

reductions in manufacturing and assembly time costs.

**Parts List**

1 axle housing with center stud
1 damper plate

2 parts total

75 % part reduction

*Figure 4-7: Redesigned Damper Device*

## Pneumatic Pressure Sensor

Shown in Figure 4-8 is a disassembled pneumatic pressure sensor consisting of seven separate parts. When pressure is increased at the lower inlet port of the base the piston is raised against the coil spring.

**Parts List**

2 screws
1 cover plate
1 helical spring
1 piston stop
1 piston
1 base

7 parts total

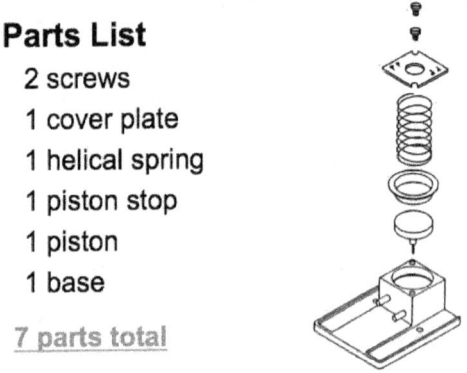

*Figure 4-8: Pneumatic Pressure Sensor*

Design for Assembly

If the pressure is high enough the piston will move upward against the piston stop opening the outlet to the higher pressure in the body. The device is capped and held together with a cover that is fastened to the base by two machine screws.

Assembly begins with the base as the base component. The piston is inserted into the cavity in the base followed by the piston stop. The coil spring is then inserted and finally the cap is placed on the base and is fastened down with two machine screws. It is observed that the assembly is all vertically down and sequentially layered satisfying that aspect of DFA guidelines for products.

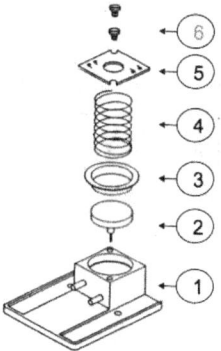

*Figure 4-9: Assembly (Pressure Sensor)*

## Part Sequence Diagram

The part sequence diagram is represented by a simple three-step process with simple vertical insertion of the piston, piston stop and coil spring into the base prior to being capped and composed by the cover and two fasteners as a final assembly.

69

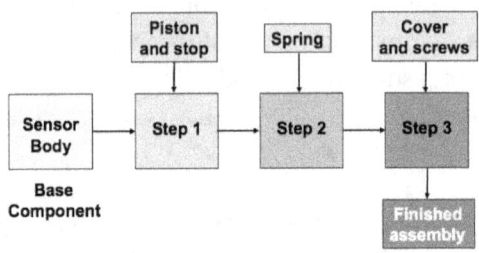

*Figure 4-10: Part Sequence Diagram (Pressure Sensor)*

## DFA System Analysis

Even though there are only two fasteners they still represent 20% of the total part count, which is quite high. The cover with two opposite asymmetric corner slots for the fasteners will require special orientation handling. The piston stop does not move, does not have to be made of a different material and does not need to be separate for assembly or disassembly; hence it may be a candidate for elimination.

- Two fasteners represent 20% of total part count – too high!
- Cover has asymmetric features – will require special orientation
- Piston stop doesn't move, doesn't need to be of a different material, doesn't need to be separate for assembly – candidate for elimination
- Spring has open ends
- Piston will be difficult to insert – no easy way to hold

*Figure 4-11: System Analysis (Pressure Sensor)*

The spring appears to have open coil ends that will tangle easily in feeding. Finally the piston will be difficult to insert, as there is no easy way to hold or guide it.

## Design Change Candidates

By forming the cap from plastic and having it snapped into the top of the base the two screws can be eliminated. This will require a grove to be machined inside the base near its top to provide sufficient quality for the composition of the new cap. The piston stop can be eliminated by combining it with the redesigned cap. Placing a stud on the top of the piston will provide a gripping point to assist orientation for its insertion. Finally the end of the coli spring can be closed to help reduce tangling.

> Eliminate screws and use snap fit on plastic circular cap (eliminates fasteners −2 parts)
>
> Combine cover cap and piston stop − (eliminates 1 part)
>
> Place stud on piston to provide guiding in assembly
>
> Close spring ends

*Figure 4-12: Design Change Candidates (Pressure Sensor)*

## Redesigned Sensor

The redesigned sensor is shown in Figure 4-13. Vertical assembly has been maintained but the part count has been reduced from seven parts to four. This is a reduction of 57% and with the elimination of the separate fasteners represents a significant improvement in ease of assembly and reduction in cost.

Design for Assembly

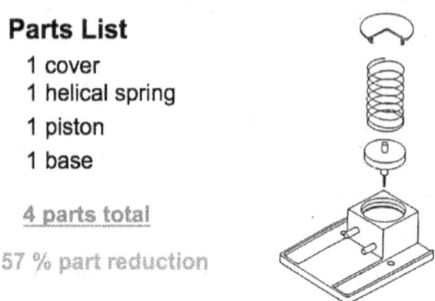

**Parts List**

1 cover
1 helical spring
1 piston
1 base

4 parts total

57 % part reduction

*Figure 4-13: Redesigned Sensor*

## Light Switch Redesign

The next several examples are from projects undertaken by students taking a course in Design for Assembly at North Carolina State University. They are presented with less detail than the previous but the evidence of ease of assembly improvement will be clearly evident.

The first example is a common residential light switch shown disassembled in Figure 4-14. It consists of some nineteen separate parts some of which are repeated. Observation of the redesign in Figure 4-15 shows a significant reduction in part count. A number of fasteners have been eliminated and several parts have either been combined into separate subassemblies or parts that combine several functions.

Design for Assembly

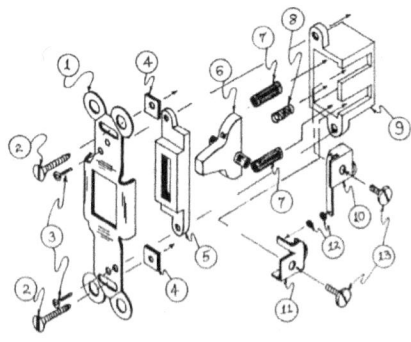

*Figure 4-14: Original Light Switch*

Part 5 is a subassembly of the electrical components and part 3 combines the effect of the toggle bumpers and toggle spring of the original design. Snap fasteners for the exterior cover also have eliminated several fasteners and the redesigned screws, part 6, combine the functions of the original wall attachment screws and retaining washers.

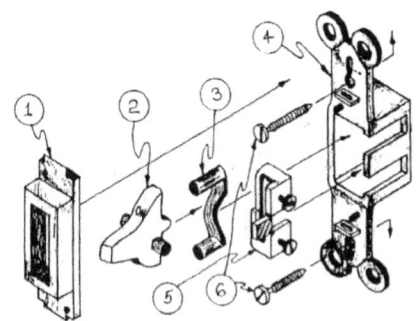

*Figure 4-15: Redesigned Light Switch*

Design for Assembly

## Plastic Pipe Cutter

Shown in Figure 4-16 is a commercial plastic pipe cutter that under went redesign. It operates by first placing the pipe into the holding jaw. The selector switch is then set to bring the cutter blade into contact with the pipe. The trigger handle is squeezed repeatedly ratcheting the cutter blade through the pipe. When the pipe is severed the release button is pressed withdrawing the blade back into the cutter body.

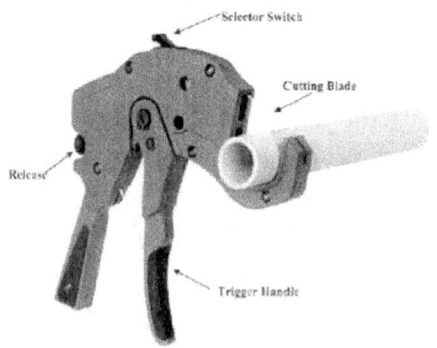

*Figure 4-16: Original Pipe Cutter*

The completely disassembled cutter is shown in Figure 4-17. The cutter consists of five separate subassemblies and accounting for the additional parts required for final assembly totals 52 separate parts.

Design for Assembly

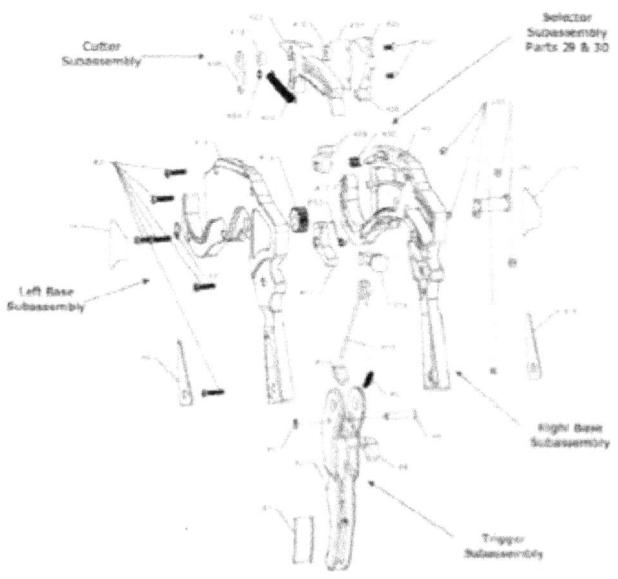

*Figure 4-17: Disassembled Cutter*

The number of fasteners in the original design was not excessive; therefore redesign strategy concentrated on combing parts while retaining functionality. This resulted in the five subassemblies being reduced to three with a significant reduction in part count. With the assemblies counted as single parts in the final assembly the total part count was reduced to 22, less than half of the original part count, another example of the impact of DFA principles.

Design for Assembly

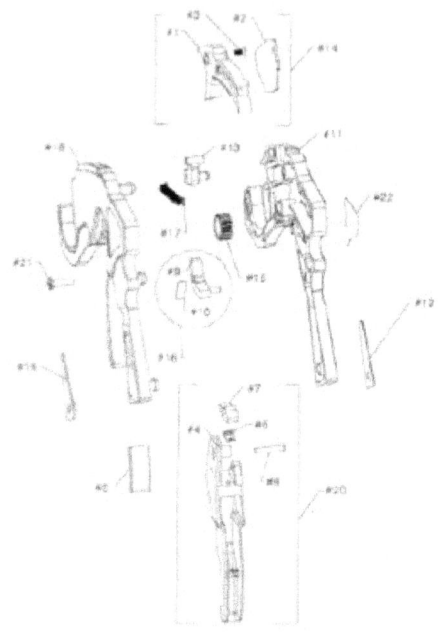

*Figure 4-18: Re designed Cutter*

The redesigned cutter is depicted in Figure 4-18. A visual comparison of the redesign to the original design demonstrates clearly the impact that the application of the DFA guidelines can make

## Copier Latch Subassembly

Depicted in Figure 4-19 is the latching subassembly from a nationally produced commercial photocopier. The manufacturer under took its redesign applying the principles of DFA to determine potential cost reduction even though the copier was already in production and distribution.

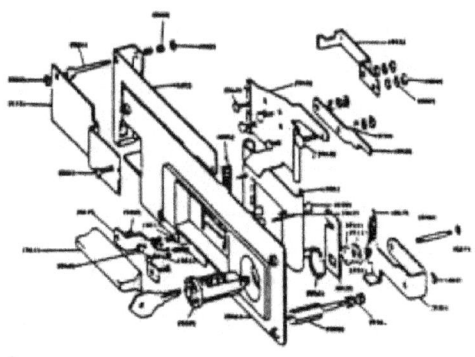

*Figure 4-19: Copier Latch Subassembly*

The result of this effort resulted in the depiction of the redesigned latching subassembly shown in Figure 4-20.

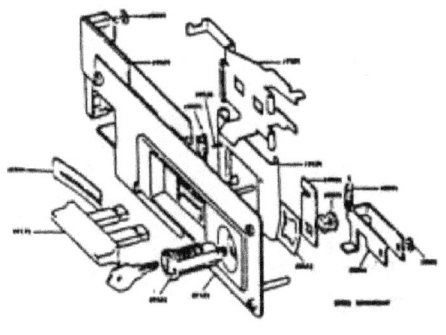

*Figure 4-20: Copier Latch Redesign*

## Design for Assembly

This redesign was obviously successful in significantly reducing the total part count and number of separate subassemblies. In addition to the reduction in part count, estimates of the assembly time and device cost are compared with those available from the original design. Both the part count and assembly time underwent very significant percentage reductions. The total cost of the subassembly was also substantially reduced. Although this represented a successful application of DFA the redesign was never implemented. The remaining product run on this copier was limited and retooling cost would be prohibitive so the redesign became nothing more than an interesting exercise that carried with it an important lesson.

| Redesign of Copier Latch | | | |
|---|---|---|---|
| | Old Design | New Design | Percent Change |
| Total Number of Parts | 62 | 17 | 73% |
| Est. Assembly Time (min.) | 6.90 | 1.48 | 79% |
| Est. Product Cost | $12.56 | $8.03 | 36% |

*Figure 4-21  DFA Redesign Impact*

The message from the copier latch subassembly exercise is simply that the application of DFA principles and design effort that will provide for ease of assembly should be applied from the very first day of product development to have its maximum effect.

# Chapter 5 – Quantifying Ease of Assembly

Chapter 5 presents a tool for quantifying numerically the improvement in ease of assembly in the redesign of an existing product. It is based on the ranking of the physical activities and tasks associated with the process of assembly.

## Basis of Quantitative Tool

This quantitative tool is based on assigning rating values for each part from 0 to 10 depending on how easy or hard it is to feed, insert and fasten into the assembly. These rating values are then combined to create a merit value for the part as it relates to ease of assembly. In addition the part redundancy criteria from Chapter 3 is applied to each part to determine whether it may be a potential candidate for elimination.

> Each part is numerically rated from 0 to 10
> (hard to easy) for following assembly events –
>   Feeding
>   Insertion
>   Fastening
>
> Event ratings for each part are combined
>   into a numerical merit for the part
>
> Redundancy criteria is applied to each
>   part to determine its potential for elimination

*Figure 5-1: Summary of Basis of Quantitative Tooling*

# Design for Assembly

## Part Merit Rating

The rating values for feeding, insertion and fastening are then combined into a single part figure of merit. The model used for this assumes that each rating component can be represented as a coordinate on the axes of a three dimensional coordinate system. The length of the vector in this three-dimensional space defined by these three coordinates is a measure of the part figure of merit. The higher each component is rated, the longer the resulting vector and the higher the figure of merit for the part.

## Geometric Interpretation

Shown in Figure 5-2 is a model of the three dimensional space used to define each part figure of merit. A rating on a scale of 0 to 10 with 10 being the best is first established for each part. These ratings are then plotted on the three axes of feeding, insertion and fastening. This is represented in the figure by the three gray coordinate vectors of feeding equal to 8, insertion equal to 7 and fastening equal to 9. The length of the combined gray vector becomes the part figure of merit defined by these three coordinates.

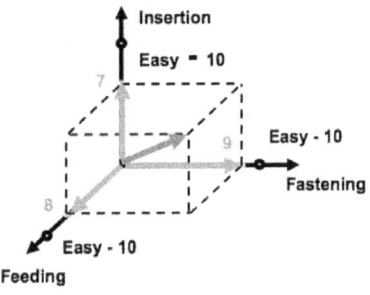

*Figure 5-2: Geometric Interpretation*

The largest possible figure of part merit would be defined by component ratings for feeding, insertion and fastening of 10 indicating that this represents the easiest or best possible value for each of these activities of assembly.

## Part Merit Calculation

The numerical value of the length of the part merit vector is calculated as the square root of the sum of the squares of the feeding, insertion and fastening rating values. This is divided by the square root of three to normalize the final part merit to a maximum value of 10. Without this normalization the value of the part merit for ratings of 10 for feeding, insertion and fastening would be 10 times the square root of three which would be inconvenient relative to the manner in which these part merits will be used.

$$FeR = \text{Feed Rating}$$
$$InR = \text{Insert Rating}$$
$$FaR = \text{Fasten Rating}$$

$$\text{Part Merit} = \frac{\sqrt{(FeR)^2 + (InR)^2 + (FaR)^2}}{\sqrt{3}}$$

*Figure 5-3: Part Merit Calculation*

## Assembly Merit

To relate the individual part merits to the ease of assembly of the entire product two complete assembly merit figures are defined. The first is the combined average merit. It is the sum of all the individual part merits divided by the sum of all the parts. Comparing this value for an original design to the same value for a redesign is an indication of how well the feeding, insertion and fastening may have been improved. The second assembly merit is the product assembly merit. This is equal to the combined average merit times the quantity one minus the sum of the redundant parts divided by the sum of all the parts. This is a measure of how well the parts that were identified as redundant in the original design were successfully eliminated in the redesign. The greater these two complete assembly merit values are between the original design and the redesign is an indication of the positive impact of the application of DFA in the redesign.

**Combined Average Merit**

$$CAM = \frac{\text{Sum of part merits}}{\text{Sum of parts}}$$

**Product Assembly Merit**

$$PAM = CAM \left\{ 1 - \frac{\text{Sum of redundant parts}}{\text{Sum of parts}} \right\}$$

*Figure 5-4: Assembly Merit Calculation*

## Feeding Ratings

The quantitative tool is based on rating values for each part from 0 to 10 depending on how easy or hard it is to feed, insert and fasten into the assembly. This subjective rating requires the assignment of a numerical value to proceed with the application of the tool. How rating values from zero to ten are chosen for a specific product and its assembly is a subjective choice based on mutual agreement of those who wish to apply the tool

Design for Assembly

and benefit from its results. One example of how this might proceed is illustrated in Figure 5-5.

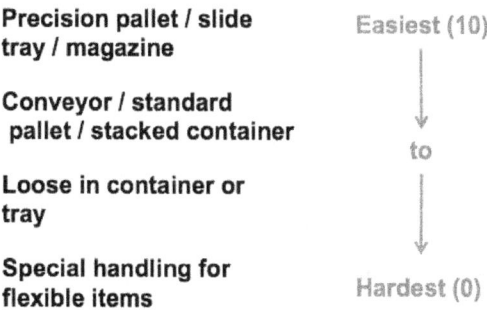

*Figure 5-5: Feeding Ratings*

For manual assembly it would be relatively easy to feed a part if it were supplied in a precision pallet, slide tray, and or magazine that already provided proper orientation and separation. The rating assigned in this instance is a ten. The feeding of flexible items that would require special handling or separation and orientation might be considered the hardest to perform and be assigned a zero. Two other intermediate possible feeding scenarios are also suggested and some rating between zero and ten would have to be assigned to them that are acceptable to the users of the tool.

## Insertion Ratings

In a similar manner numerical rating values need to be assigned for part insertions of different degrees of complexity. In this instance vertically down from the top is assigned a ten while vertically up from the bottom is given a rating of zero. Other possible examples of insertion of different degrees of complexity are listed

83

requiring some numerical rating assignment between 10 and zero. It is important to remember that all numerical rating assignments to specific feeding, insertion and fastening defined activities are subjective and are meaningful only to those who agree that they have meaning in their application to the assembly of a specific product.

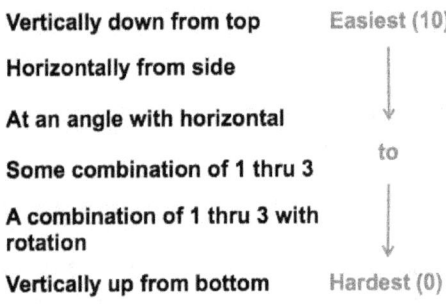

*Figure 5-6: Insertion Ratings*

## Fastener Time Comparisons

In some instances assistance can be provided to the rating process by adapting information available from published sources. For example the table of assembly time comparisons in Figure 5-7 could be used to establish rating values for the fastening function. It might start with no fastening required as being a 10 and then running down this list with screw/ washer/nut being 0 and all the other fastening possibilities distributed in between in accordance with their comparative fastening times.

Design for Assembly

| Fasteners Assembly Times | |
|---|---|
| Method | Assembly Time (secs) |
| Snap Fit | 4.1 |
| Press Fit | 7.3 |
| Integral Screw Fasteners | 11.5 |
| Rivets(4) | 36.1 |
| Machine Screw (4) | 40.5 |
| Screw/Washer/Nut (4) | 73.8 |

*Figure 5-7: Fastener Assembly Timings*

## Pneumatic Pressure Sensor

The assembly-quantifying tool will now be applied in detail to the pneumatic pressure sensor and redesign from Chapter 4. The original design and redesign are repeated here for reference.

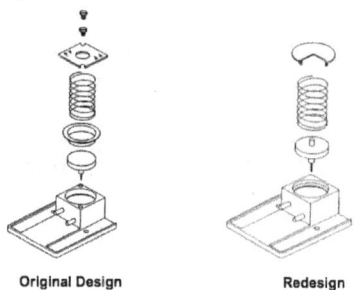

*Figure 5-8: Pneumatic Pressure Sensor*

## Original Pneumatic Sensor Rating

To implement the tool a table (Figure 5-9) is created with a series of columns for the part name and the number of parts followed by the part's feeding, insertion and fastening ratings. The next column is the value of the calculated part merit. The final column is

85

used to indicate whether it is a potentially redundant part. A zero designates it is not a redundant part and a one is assigned if it is possibly redundant.

| Original Pneumatic Piston | | | | | | |
|---|---|---|---|---|---|---|
| Part Name | No. | Feeding | Insertion | Fastening | Part Merit | Redundant Part |
| Base | 1 | 10 | 10 | 10 | 10.00 | 0 |
| Piston | | | | | | |
| Helical Spring | | | | | | |
| Piston Stop | | | | | | |
| Cover | | | | | | |
| Screw | | | | | | |
| Screw | | | | | | |
| SUM | | | | | | |
| | | | | | | |
| | | | | CAM | | |
| | | | | PAM | | |

*Figure 5-9: Base Part Merit*

It will be assumed that the sensor will be assembled manually. The parts will be listed in their insertion sequence. The first component is the base that serves as the base component. Assuming it is being delivered in some form of precision tray or magazine there will be no feeding problem. It isn't being inserted and it does not need to be fastened, hence all three rating values are given a 10. This results in a part merit of ten and the base certainly isn't a redundant part.

The piston is the next part in sequence. Feeding again should not be difficult but insertion will pose problems because the piston must be gripped manually on it sides making it cumbersome to slip it into the bore of the base. It is rated 7 for this action. The piston requires no separate fastening so this rating will be a 10. The calculated part merit is 9.11 and it is certainly not a redundant part. (See Figure 5-10)

Design for Assembly

| Original Pneumatic Piston | | | | | | Redundant Part |
|---|---|---|---|---|---|---|
| Part Name | No. | Feeding | Insertion | Fastening | Part Merit | |
| Base | 1 | 10 | 10 | 10 | 10.00 | 0 |
| Piston | 1 | 10 | 7 | 10 | 9.11 | 0 |
| Helical Spring | | | | | | |
| Piston Stop | | | | | | |
| Cover | | | | | | |
| Screw | | | | | | |
| Screw | | | | | | |
| SUM | | | | | | 0 |
| | | | | CAM | | |
| | | | | PAM | | |

*Figure 5-10: Piston Part Merit*

The spring is the next part to be added. If delivered in bulk the open ends will create tangling making feeding difficult resulting in a rating of 3. Once a single spring is available the insertion in the base will be easy and no fastening is required so both of these ratings are given a 10. The part merit for the spring is 8.35. It is also required for function so a zero is entered in the redundant part column. (See Figure 5-11)

| Original Pneumatic Piston | | | | | | Redundant Part |
|---|---|---|---|---|---|---|
| Part Name | No. | Feeding | Insertion | Fastening | Part Merit | |
| Base | 1 | 10 | 10 | 10 | 10.00 | 0 |
| Piston | 1 | 10 | 7 | 10 | 9.11 | 0 |
| Helical Spring | 1 | 3 | 10 | 10 | 8.35 | 0 |
| Piston Stop | | | | | | |
| Cover | | | | | | |
| Screw | | | | | | |
| Screw | | | | | | |
| SUM | | | | | | 0 |
| | | | | CAM | | |
| | | | | PAM | | |

*Figure 5-11: Spring Part Merit*

The piston stop is now added. Its symmetry will make it easy to orient and feed, it simply inserts vertically into the bore and no fastening is required. Ratings of 10 are assigned to all three activities resulting in a part merit of 10. Since it can be eliminated by

# Design for Assembly

redesign of the cover it is given a one in the final column. (See Figure 5-12)

| Original Pneumatic Piston | | | | | | |
|---|---|---|---|---|---|---|
| Part Name | No. | Feeding | Insertion | Fastening | Part Merit | Redundant Part |
| Base | 1 | 10 | 10 | 10 | 10.00 | 0 |
| Piston | 1 | 10 | 7 | 10 | 9.11 | 0 |
| Helical Spring | 1 | 3 | 10 | 10 | 8.35 | 0 |
| Piston Stop | 1 | 10 | 10 | 10 | 10.00 | 1 |
| Cover | | | | | | |
| Screw | | | | | | |
| Screw | | | | | | |
| SUM | | | | | | 1 |
| | | | | CAM | | |
| | | | | PAM | | |

*Figure 5-12: Piston Stop Part Merit*

With its asymmetric exterior geometric features the cover will require special handling and orientation so its feed rating is only given a 3. Once fed its insertion is simple and is ranted a 10. However, after being placed on top of the sensor body it must be held down to provide some initial compression for the spring so this rating is reduced from a 10 to 9. The final part value is 7.96. A cover for the device is required even though it may be redesigned. It is not considered redundant. (See Figure 5-13)

| Original Pneumatic Piston | | | | | | |
|---|---|---|---|---|---|---|
| Part Name | No. | Feeding | Insertion | Fastening | Part Merit | Redundant Part |
| Base | 1 | 10 | 10 | 10 | 10.00 | 0 |
| Piston | 1 | 10 | 7 | 10 | 9.11 | 0 |
| Helical Spring | 1 | 3 | 10 | 10 | 8.35 | 0 |
| Piston Stop | 1 | 10 | 10 | 10 | 10.00 | 1 |
| Cover | 1 | 3 | 10 | 9 | 7.96 | 0 |
| Screw | | | | | | |
| Screw | | | | | | |
| SUM | | | | | | |
| | | | | CAM | | |
| | | | | PAM | | |

*Figure 5-13: Cover Part Merit*

# Design for Assembly

If fed in bulk the screws will also provide some handling and fastening problems. Its feeding is assigned a rating of 3. Their insertion is vertically down and is straightforward so a rating of 10 is appropriate. Since they will require tooling to be threaded and torqued a rating of 3 is assigned resulting in a part merit of 6.27. The screws are listed separately so that the total part count is correct for the calculation of the combined average merit and the product assembly merit. A common practice is to list all fasteners as being potentially redundant to help drive their elimination where possible in redesign. (See Figure 5-14)

| Part Name | No. | Feeding | Insertion | Fastening | Part Merit | Redundant Part |
|---|---|---|---|---|---|---|
| Base | 1 | 10 | 10 | 10 | 10.00 | 0 |
| Piston | 1 | 10 | 7 | 10 | 9.11 | 0 |
| Helical Spring | 1 | 3 | 10 | 10 | 8.35 | 0 |
| Piston Stop | 1 | 10 | 10 | 10 | 10.00 | 1 |
| Cover | 1 | 3 | 10 | 9 | 7.96 | 0 |
| Screw | 1 | 3 | 10 | 3 | 8.27 | 1 |
| Screw | 1 | 3 | 10 | 3 | 6.27 | 1 |
| SUM | | | | | | |
| | | | | CAM | | |
| | | | | PAM | | |

Original Pneumatic Piston

*Figure 5-14: Screw Part Merit*

The part merit numbers are now summed and divided by the total part count to give a combined average merit of 8.28. With three parts indicated as potentially redundant the product assembly merit is calculated as 4.73 (Figure 5-15). There are no absolute passing scores for these values. Their significance lies in how close they are to what would be a perfect score of 10 and how large the difference is between the two values. These two considerations indicate where emphasis should be placed in the redesign

## Design for Assembly

**Original Pneumatic Piston**

| Part Name | No. | Feeding | Insertion | Fastening | Part Merit | Redundant Part |
|---|---|---|---|---|---|---|
| Base | 1 | 10 | 10 | 10 | 10.00 | 0 |
| Piston | 1 | 10 | 7 | 10 | 9.11 | 0 |
| Helical Spring | 1 | 3 | 10 | 10 | 8.35 | 0 |
| Piston Stop | 1 | 10 | 10 | 10 | 10.00 | 1 |
| Cover | 1 | 3 | 10 | 9 | 7.96 | 0 |
| Screw | 1 | 3 | 10 | 3 | 6.27 | 1 |
| Screw | 1 | 3 | 10 | 3 | 6.27 | 1 |
| SUM | 7 | | | | 57.96 | 3 |
| | | | | CAM | 8.28 | |
| | | | | PAM | 4.73 | |

*Figure 5-15: CAM & PAM Score*

## Redesigned Pneumatic Sensor

The merit table is now recreated for evaluating the numerical ease of assembly for the redesigned sensor. No changes have been made in the sensor base so its rating and merit figures remain the same and it cannot be considered a redundant part. (Figure 5-16)

**Redesigned Pneumatic Piston**

| Part Name | No. | Feeding | Insertion | Fastening | Part Merit | Redundant Part |
|---|---|---|---|---|---|---|
| Block | 1 | 10 | 10 | 10 | 10.00 | 0 |
| Piston | | | | | | |
| Helical Spring | | | | | | |
| Cover | | | | | | |
| SUM | | | | | | |
| | | | | CAM | | |
| | | | | PAM | | |

*Figure 5-16: New Block Part Merit*

In the redesign a stub was added to the center of the top of the piston to provide easier gripping to assist in insertion. To account for this the insertion rating has been increased from 7 to 9. The other ratings remain the same and the part merit increases to 9.68. It also is a required part so zero is repeated in the redundant part column. (Figure 5-17)

### Redesigned Pneumatic Piston

| Part Name | No. | Feeding | Insertion | Fastening | Part Merit | Redundant Part |
|---|---|---|---|---|---|---|
| Block | 1 | 10 | 10 | 10 | 10.00 | 0 |
| Piston | 1 | 10 | 9 | 10 | 9.68 | 0 |
| Helical Spring | | | | | | |
| Cover | | | | | | |
| SUM | | | | | | |
| | | | | CAM | | |
| | | | | PAM | | |

*Figure 5-17: New Piston Part Merit*

The ends of the coil spring have been closed which improves its feeding from 3 to 8. With its other ratings remaining the same the part merit value is now 9.38. It is not a redundant part either. (Figure 5-18)

### Redesigned Pneumatic Piston

| Part Name | No. | Feeding | Insertion | Fastening | Part Merit | Redundant Part |
|---|---|---|---|---|---|---|
| Block | 1 | 10 | 10 | 10 | 10.00 | 0 |
| Piston | 1 | 10 | 9 | 10 | 9.68 | 0 |
| Helical Spring | 1 | 8 | 10 | 10 | 9.38 | 0 |
| Cover | | | | | | |
| SUM | | | | | | |
| | | | | CAM | | |
| | | | | PAM | | |

*Figure 5-18: New Spring Part Merit*

The original cover that is now a circular plastic cap that snaps into the bore of the body has eliminated the piston stop and the two screws. This new cap is certainly easy to feed and insert so those ratings are given a 10. Since it needs to be snapped into the bore requiring some force its fastening ranking is given a 9.

The part merit value is 9.68. It too is a required part. (Figure 5-19)

| Redesigned Pneumatic Piston | | | | | | |
|---|---|---|---|---|---|---|
| Part Name | No. | Feeding | Insertion | Fastening | Part Merit | Redundant Part |
| Block | 1 | 10 | 10 | 10 | 10.00 | 0 |
| Piston | 1 | 10 | 9 | 10 | 9.68 | 0 |
| Helical Spring | 1 | 8 | 10 | 10 | 9.38 | 0 |
| Cover | 1 | 10 | 10 | 9 | 9.68 | 0 |
| SUM | | | | | | |
| | | | | CAM | | |
| | | | | PAM | | |

*Figure 5-19: New Cover Part Merit*

The total part number has been reduced to four from seven and the combined average merit value is increased to 9.68 from 8.28 (figure 5-20). The product assembly merit is the same as the combined average merit since all redundant parts have been eliminated. Comparison of these numerical final merit values between the original design and the redesign gives a more objective view of how ease of assembly has been improved.

| Redesigned Pneumatic Piston | | | | | | |
|---|---|---|---|---|---|---|
| Part Name | No. | Feeding | Insertion | Fastening | Part Merit | Redundant Part |
| Block | 1 | 10 | 10 | 10 | 10.00 | 0 |
| Piston | 1 | 10 | 9 | 10 | 9.68 | 0 |
| Helical Spring | 1 | 8 | 10 | 10 | 9.38 | 0 |
| Cover | 1 | 10 | 10 | 9 | 9.68 | 0 |
| SUM | 4 | | | | 38.74 | 0 |
| | | | | CAM | 9.68 | |
| | | | | PAM | 9.68 | |

*Figure 5-20: New CAM & PAM*

Design for Assembly

## Pipe Cutter Redesign Rating

The pipe cutter from Chapter 4 is shown In Figure 5-21 with both its original and redesign part assembly drawings. It was analyzed using the ease of assembly quantifying tool.

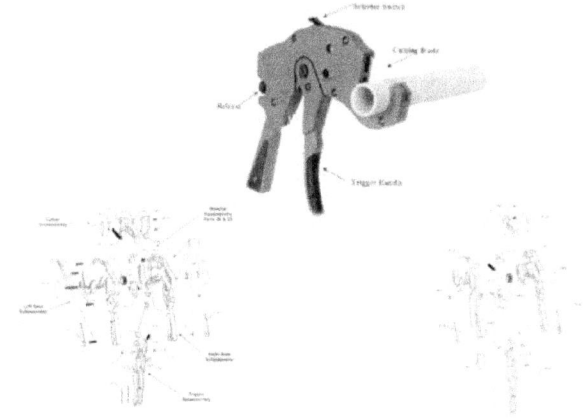

*Figure 5-21: Pipe Cutter Redesign*

## Pipe Cutter CAM and PAM

Presented here without the details of how they were determined are the combined average merit and product assembly merit for both the original and redesigned pipe cutter. Recall that the redesign strategy was to combine parts to eliminate subassemblies while maintaining functionality. In the first tables in Figure 5-22 the individual subassemblies are treated as separate products with their own combined average merit and product assembly merit. In determining these values for the final product assembly each subassembly is treated as a separate part using the combined average merit from their individual analyses as a part merit. The last table lists the percentage improvement for the final subassemblies and the final product. The ease of assembly as quantified by this tool for the pipe cutter is significantly improved by a value of almost 60 %.

| Original Assembly | CAM | PAM | Redesigned Assembly | CAM | PAM | % CAM Increase | % PAM Increase |
|---|---|---|---|---|---|---|---|
| Trigger | 5.21 | 2.23 | Trigger | 8.47 | 6.77 | 62.57 | 203.59 |
| Right Base | 4.95 | 1.98 | lock | 9.18 | 9.18 | NA | NA |
| Left Base | 7.09 | 2.36 | | | | NA | NA |
| Cutter | 5.55 | 3.09 | Cutter | 9.23 | 9.23 | 66.31 | 198.71 |
| Selector | 7.48 | 3.74 | | | | NA | NA |
| Final Assembly | 6.03 | 4.66 | Final Assembly | 8.92 | 7.43 | 47.93 | 59.44 |

*Figure 5-22: CAM & PAM Scores for Pipe Cutter*

## Comparison of Results

A comparison of the combined average merit and the product assembly merit of an original design indicates where the application of DFA in the redesign can have the greatest impact. If the result is a high combined average merit but low product assembly merit elimination of redundant parts should be emphasized. If the combined average merit is low but the product assembly merit is close to this value then apparently there are few redundant part but the feeding, insertion and fastening needs to be improved.

1. **High CAM   Low PAM**
   Eliminate redundant parts
2. **Low CAM   High PAM**
   Improve feeding, insertion, fastening
3. **Low CAM   Low PAM**
   Improve feeding, insertion, fastening and eliminate redundant part

*Figure 5-23: Comparison Summary*

# Design for Assembly

If both the combined average merit and the product assembly merit are low then redesign needs to address both feeding, insertion and fastening as well as the elimination of redundant parts.

It should be kept in mind that the calculated merit values for individual parts and the entire assembly are based on subjective judgments. These values only have meaning if all who will use this procedure to compare an original to a redesigned product agree on the numerical values used to rate the individual assembly actions. The procedure can also be used to quantify ease of assembly in the development of a new product by comparing the results to that of an "ideal" product rank at 10 for both CAM, the combined part merit, and the product assembly merit, PAM.

# Design for Assembly